CHRISTIE JOHN GEANKOPLIS

University of Minnesota

Transport Processes and Separation Process Principles
(Includes Unit Operations)

FOURTH EDITION

PRENTICE
HALL
PTR

PRENTICE HALL
Professional Technical Reference
Upper Saddle River, NJ 07458
www.phptr.com

ISBN 0-13-101367-X

90000

9 790131 013673

Library of Congress Cataloging-in-Publication Data

A CIP record for this book can be obtained from the Library of Congress.

Editorial Production/Composition: *G & S Typesetters, Inc.*
Cover Director: *Jerry Votta*
Art Director: *Gail Cocker-Bogusz*
Manufacturing Buyer: *Maura Zaldivar*
Publisher: *Bernard Goodwin*
Editorial Assistant: *Michelle Vincenti*
Marketing Manager: *Dan DePasquale*

PRENTICE
HALL
PTR
© 2003 by Pearson Education, Inc.
Publishing as Prentice Hall Professional Technical Reference
Upper Saddle River, New Jersey 07458

Prentice Hall books are widely used by corporations and government agencies for training, marketing, and resale.

For information regarding corporate and government bulk discounts please contact:
Corporate and Government Sales (800) 382-3419 or corpsales@pearsontechgroup.com

Other company and product names mentioned herein are the trademarks or registered trademarks of their respective owners.

ISBN 0-13-101367-X

Text printed in the United States at Courier Westford in Westford, Massachusetts.

15th Printing   December 2009

Pearson Education LTD.
Pearson Education Australia PTY, Limited
Pearson Education Singapore, Pte. Ltd.
Pearson Education North Asia Ltd.
Pearson Education Canada, Ltd.
Pearson Educación de Mexico, S.A. de C.V·
Pearson Education—Japan
Pearson Education Malaysia, Pte. Ltd.

Dedicated to the memory of my beloved mother, Helen,
for her love and encouragement

About Prentice Hall Professional Technical Reference

With origins reaching back to the industry's first computer science publishing program in the 1960s, and formally launched as its own imprint in 1986, Prentice Hall Professional Technical Reference (PH PTR) has developed into the leading provider of technical books in the world today. Our editors now publish over 200 books annually, authored by leaders in the fields of computing, engineering, and business.

Our roots are firmly planted in the soil that gave rise to the technical revolution. Our bookshelf contains many of the industry's computing and engineering classics: Kernighan and Ritchie's *C Programming Language*, Nemeth's *UNIX System Adminstration Handbook*, Horstmann's *Core Java*, and Johnson's *High-Speed Digital Design*.

PH PTR acknowledges its auspicious beginnings while it looks to the future for inspiration. We continue to evolve and break new ground in publishing by providing today's professionals with tomorrow's solutions.

PRENTICE
HALL
PTR

Contents

Preface

The title of this text has been changed from *Transport Processes and Unit Operations* to *Transport Processes and Separation Process Principles (Includes Unit Operations)*. This was done because the term "unit operations" has been largely superseded by the term "separation processes," which better reflects the modern nomenclature being used.

In this fourth edition, the main objectives and the format of the third edition remain the same. The sections on momentum transfer have been greatly expanded, especially the sections on fluidized beds, flow meters, mixing, and non-Newtonian fluids. Material has been added to the chapters on mass transfer. The chapters on absorption, distillation, and liquid–liquid extraction have also been enlarged. More new material has been added to the sections on ion exchange and crystallization. The chapter on membrane separation processes has been greatly expanded, especially for gas-membrane theory.

The field of chemical engineering involved with physical and physical–chemical changes of inorganic and organic materials and, to some extent, biological materials is overlapping more and more with the other process-engineering fields of ceramic engineering, process metallurgy, agricultural food engineering, wastewater-treatment (civil) engineering, and bio-engineering. The principles of momentum, heat, and mass transport and the separation processes are widely used in these processing fields.

The principles of momentum transfer and heat transfer have been taught to all engineers. The study of mass transfer has been limited primarily to chemical engineers. However, engineers in other fields have become more interested in mass transfer in gases, liquids, and solids.

Since chemical and other engineering students must study so many topics today, a more unified introduction to the transport processes of momentum, heat, and mass transfer and to the applications of separation processes is provided. In this text the principles of the transport processes are covered first, and then the separation processes (unit operations). To accomplish this, the text is divided into two main parts.

PART 1: Transport Processes: Momentum, Heat, and Mass

This part, dealing with fundamental principles, includes the following chapters: 1. Introduction to Engineering Principles and Units; 2. Principles of Momentum Transfer and Overall Balances; 3. Principles of Momentum Transfer and Applications; 4. Principles of Steady-State

Heat Transfer; 5. Principles of Unsteady-State Heat Transfer; 6. Principles of Mass Transfer; and 7. Principles of Unsteady-State and Convective Mass Transfer.

PART 2: Separation Process Principles (Includes Unit Operations)

This part, dealing with applications, covers the following separation processes: 8. Evaporation; 9. Drying of Process Materials; 10. Stage and Continuous Gas–Liquid Separation Processes (humidification, absorption); 11. Vapor–Liquid Separation Processes (distillation); 12. Liquid–Liquid and Fluid–Solid Separation Processes (adsorption, ion exchange, extraction, leaching, crystallization); 13. Membrane Separation Processes (dialysis, gas separation, reverse osmosis, ultrafiltration, microfiltration); 14. Mechanical–Physical Separation Processes (filtration, settling, centrifugal separation, mechanical size reduction).

In Chapter 1 elementary principles of mathematical and graphical methods, laws of chemistry and physics, material balances, and heat balances are reviewed. Many readers, especially chemical engineers, may be familiar with most of these principles and may omit all or parts of this chapter.

A few topics, primarily those concerned with the processing of biological materials, may be omitted at the discretion of the reader or instructor; these include Sections 5.5, 6.4, 8.7, 9.11, and 9.12. Over 240 example or sample problems and over 550 homework problems on all topics are included in the text. Some of the homework problems involve biological systems, for those readers who are especially interested in that area.

This text may be used for a course of study following any of the following five suggested plans. In all plans, Chapter 1 may or may not be included.

1. *Study of transport processes of momentum, heat, and mass and separation processes.* In this plan, most of the entire text, covering the principles of the transport processes in Part 1 and the separation processes in Part 2, is covered. This plan would be applicable primarily to chemical engineering as well as to other process-engineering fields in a one-and-one-half-year course of study at the junior and/or senior level.

2. *Study of transport processes of momentum, heat, and mass and selected separation processes.* Only the elementary sections of Part 1 (the principles chapters—2, 3, 4, 5, 6, and 7) are covered, plus selected separation-processes topics in Part 2 applicable to a particular field, in a two-semester or three-quarter course. Students in environmental engineering, food process engineering, and process metallurgy could follow this plan.

3. *Study of transport processes of momentum, heat, and mass.* The purpose of this plan in a two-quarter or two-semester course is to obtain a basic understanding of the transport processes of momentum, heat, and mass transfer. This involves studying sections of the principles chapters—2, 3, 4, 5, 6, and 7 in Part 1—and omitting Part 2, the applied chapters on separation processes.

4. *Study of separations processes.* If the reader has had courses in the transport processes of momentum, heat, and mass, Chapters 2–7 can be omitted and only the separation processes chapters in Part 2 studied in a one-semester or two-quarter course. This plan could be used by chemical and certain other engineers.

5. *Study of mass transfer.* For those such as chemical or mechanical engineers who have had momentum and heat transfer, or those who desire only a background in mass transfer in a one-quarter or one-semester course, Chapters 6, 7, and 10 would be covered. Chapters 9, 11, 12, and 13 might be covered optionally, depending on the needs of the reader.

Different schools and instructors differ on the use of computers in engineering courses. All of the equations and homework problems in this text can be solved by using ordinary hand-held computers. However, more complicated problems involving numerical integration, finite-difference calculations, steady- and unsteady-state two-dimensional diffusion and conduction, and so on, can easily be solved with a computer using spreadsheets. Almost all undergraduate students are proficient in their use.

The SI (Système International d'Unités) system of units has been adopted by the scientific community. Because of this, the SI system of units has been adopted in this text for use in the equations, example problems, and homework problems. However, the most important equations derived in the text are also given in a dual set of units, SI and English, when different. Many example and homework problems are also given using English units.

Christie John Geankoplis

PART 1

Transport Processes: Momentum, Heat, and Mass

CHAPTER 1

Introduction to Engineering Principles and Units

1.1 CLASSIFICATION OF TRANSPORT PROCESSES AND SEPARATION PROCESSES (UNIT OPERATIONS)

1.1A Introduction

In the chemical and other physical processing industries and the food and biological processing industries, many similarities exist in the manner in which the entering feed materials are modified or processed into final materials of chemical and biological products. We can take these seemingly different chemical, physical, or biological processes and break them down into a series of separate and distinct steps that were originally called *unit operations*. However, the term "unit operations" has largely been superseded by the more modern and descriptive term "separation processes." These *separation processes* are common to all types of diverse process industries.

For example, the separation process *distillation* is used to purify or separate alcohol in the beverage industry and hydrocarbons in the petroleum industry. Drying of grain and other foods is similar to drying of lumber, filtered precipitates, and wool. The separation process *absorption* occurs in absorption of oxygen from air in a fermentation process or in a sewage treatment plant and in absorption of hydrogen gas in a process for liquid hydrogenation of oil. Evaporation of salt solutions in the chemical industry is similar to evaporation of sugar solutions in the food industry. Settling and sedimentation of suspended solids in the sewage industry and the mining industry are similar. Flow of liquid hydrocarbons in the petroleum refinery and flow of milk in a dairy plant are carried out in a similar fashion.

Many of these separation processes have certain fundamental and basic principles or mechanisms in common. For example, the mechanism of diffusion or mass transfer occurs in drying, membrane separation, absorption, distillation, and crystallization. Heat transfer occurs in drying, distillation, evaporation, and so on. The following classification of a more fundamental nature is often made, according to transfer or transport processes.

1.1B Fundamental Transport Processes

1. Momentum transfer. This is concerned with the transfer of momentum which occurs in moving media, such as in the separation processes of fluid flow, sedimentation, mixing, and filtration.

2. Heat transfer. In this fundamental process, we are concerned with the transfer of heat from one place to another; it occurs in the separation processes of drying, evaporation, distillation, and others.

3. Mass transfer. Here mass is being transferred from one phase to another distinct phase; the basic mechanism is the same whether the phases are gas, solid, or liquid. This includes distillation, absorption, liquid–liquid extraction, membrane separation, adsorption, crystallization, and leaching.

1.1C Classification of Separation Processes

The separation processes deal mainly with the transfer and change of energy and the transfer and change of materials, primarily by physical means but also by physical–chemical means. The important separation processes, which can be combined in various sequences in a process and which are covered in this text, are described next.

1. Evaporation. This refers to the evaporation of a volatile solvent such as water from a nonvolatile solute such as salt or any other material in solution.

2. Drying. In this operation volatile liquids, usually water, are removed from solid materials.

3. Distillation. This is an operation whereby components of a liquid mixture are separated by boiling because of their differences in vapor pressure.

4. Absorption. In this process a component is removed from a gas stream by treatment with a liquid.

5. Membrane separation. This process involves the separation of a solute from a fluid by diffusion of this solute from a liquid or gas through a semipermeable membrane barrier to another fluid.

6. Liquid–liquid extraction. In this case a solute in a liquid solution is removed by contacting with another liquid solvent that is relatively immiscible with the solution.

7. Adsorption. In this process a component of a gas or liquid stream is removed and adsorbed by a solid adsorbent.

8. Ion exchange. Certain ions in solution are removed from a liquid by an ion-exchange solid.

9. Liquid–solid leaching. This involves treating a finely divided solid with a liquid that dissolves out and removes a solute contained in the solid.

10. Crystallization. This concerns the removal of a solute such as a salt from a solution by precipitating the solute from the solution.

11. Mechanical–physical separations. These involve separation of solids, liquids, or gases by mechanical means, such as filtration, settling, centrifugation, and size reduction.

1.1D Arrangement in Parts 1 and 2

This text is arranged in two parts:

Part 1: Transport Processes: Momentum, Heat, and Mass. These fundamental principles are covered extensively in Chapters 1 through 7 in order to provide the basis for study of separation processes in Part 2 of this text.

Part 2: Separation Process Principles (Includes Unit Operations). The various separation processes and their applications to process areas are studied in Part 2 of this text.

There are a number of elementary engineering principles, mathematical techniques, and laws of physics and chemistry that are basic to a study of the principles of momentum, heat, and mass transfer and the separation processes. These are reviewed for the reader in this first chapter. Some readers, especially chemical engineers, agricultural engineers, civil engineers, and chemists, may be familiar with many of these principles and techniques and may wish to omit all or parts of this chapter.

Homework problems at the end of each chapter are arranged in different sections, each corresponding to the number of a given section in the chapter.

1.2 SI SYSTEM OF BASIC UNITS USED IN THIS TEXT AND OTHER SYSTEMS

There are three main systems of basic units employed at present in engineering and science. The first and most important of these is the *SI* (Système International d'Unités) *system,* which has as its three basic units the meter (m), the kilogram (kg), and the second (s). The others are the English foot (ft)–pound (lb)–second (s), or *English system* and the centimeter (cm)–gram (g)–second (s), or *cgs system.*

At present the SI system has been adopted officially for use exclusively in engineering and science, but the older English and cgs systems will still be used for some time. Much of the physical and chemical data and empirical equations are given in these latter two systems. Hence, the engineer not only should be proficient in the SI system but must also be able to use the other two systems to a limited extent.

1.2A SI System of Units

The basic quantities used in the SI system are as follows: the unit of length is the meter (m); the unit of time is the second (s); the unit of mass is the kilogram (kg); the unit of temperature is the kelvin (K); and the unit of an element is the kilogram mole (kg mol). The other standard units are derived from these basic quantities.

The basic unit of force is the newton (N), defined as

$$1 \text{ newton (N)} = 1 \text{ kg} \cdot \text{m/s}^2$$

The basic unit of work, energy, or heat is the newton-meter, or joule (J).

$$1 \text{ joule (J)} = 1 \text{ newton} \cdot \text{m (N} \cdot \text{m)} = 1 \text{ kg} \cdot \text{m}^2/\text{s}^2$$

Power is measured in joules/s or watts (W).

$$1 \text{ joule/s (J/s)} = 1 \text{ watt (W)}$$

The unit of pressure is the newton/m^2 or pascal (Pa).

$$1 \text{ newton/m}^2 \text{ (N/m}^2) = 1 \text{ pascal (Pa)}$$

[Pressure in atmospheres (atm) is not a standard SI unit but is being used during the transition period.] The standard acceleration of gravity is defined as

$$1 \text{ g} = 9.80665 \text{ m/s}^2$$

A few of the standard prefixes for multiples of the basic units are as follows: giga (G) = 10^9, mega (M) = 10^6, kilo (k) = 10^3, centi (c) = 10^{-2}, milli (m) = 10^{-3}, micro (μ) = 10^{-6}, and nano (n) = 10^{-9}. The prefix c is not a preferred prefix.

Temperatures are defined in kelvin (K) as the preferred unit in the SI system. However, in practice, wide use is made of the degree Celsius (°C) scale, which is defined by

$$t°C = T(K) - 273.15$$

Note that 1°C = 1 K and that in the case of temperature difference,

$$\Delta t°C = \Delta T \text{ K}$$

The standard preferred unit of time is the second (s), but time can be in nondecimal units of minutes (min), hours (h), or days (d).

1.2B   CGS System of Units

The cgs system is related to the SI system as follows:

$$1 \text{ g mass (g)} = 1 \times 10^{-3} \text{ kg mass (kg)}$$

$$1 \text{ cm} = 1 \times 10^{-2} \text{ m}$$

$$1 \text{ dyne (dyn)} = 1 \text{ g} \cdot \text{cm/s}^2 = 1 \times 10^{-5} \text{ newton (N)}$$

$$1 \text{ erg} = 1 \text{ dyn} \cdot \text{cm} = 1 \times 10^{-7} \text{ joule (J)}$$

The standard acceleration of gravity is

$$g = 980.665 \text{ cm/s}^2$$

1.2C   English fps System of Units

The English system is related to the SI system as follows:

$$1 \text{ lb mass (lb}_m) = 0.45359 \text{ kg}$$

$$1 \text{ ft} = 0.30480 \text{ m}$$

$$1 \text{ lb force (lb}_f) = 4.4482 \text{ newton (N)}$$

$$1 \text{ ft} \cdot \text{lb}_f = 1.35582 \text{ newton} \cdot \text{m (N} \cdot \text{m)} = 1.35582 \text{ joules (J)}$$

$$1 \text{ psia} = 6.89476 \times 10^3 \text{ newton/m}^2 \text{ (N/m}^2)$$

$$1.8°F = 1 \text{ K} = 1°C \text{ (centigrade or Celsius)}$$

$$g = 32.174 \text{ ft/s}^2$$

The proportionality factor for Newton's law is

$$g_c = 32.174 \text{ ft} \cdot \text{lb}_m/\text{lb}_f \cdot \text{s}^2$$

The factor g_c in SI units and cgs units is 1.0 and is omitted.

In Appendix A.1, convenient conversion factors for all three systems are tabulated. Further discussions and use of these relationships are given in various sections of the text.

This text uses the SI system as the primary set of units in the equations, sample problems, and homework problems. However, the important equations derived in the text are given in a dual set of units, SI and English, when these equations differ. Some example problems and homework problems are also given using English units. In some cases, intermediate steps and/or answers in example problems are also stated in English units.

1.2D Dimensionally Homogeneous Equations and Consistent Units

A dimensionally homogeneous equation is one in which all the terms have the same units. These units can be the base units or derived ones (for example, $\text{kg/s}^2 \cdot \text{m}$ or Pa). Such an equation can be used with any system of units provided that the same base or derived units are used throughout the equation. No conversion factors are needed when consistent units are used.

The reader should be careful in using any equation and always check it for dimensional homogeneity. To do this, a system of units (SI, English, etc.) is first selected. Then units are substituted for each term in the equation and like units in each term canceled out.

1.3 METHODS OF EXPRESSING TEMPERATURES AND COMPOSITIONS

1.3A Temperature

There are two temperature scales in common use in the chemical and biological industries. These are degrees Fahrenheit (abbreviated °F) and Celsius (°C). It is often necessary to convert from one scale to the other. Both use the freezing point and boiling point of water at 1 atmosphere pressure as base points. Often temperatures are expressed as absolute degrees K (SI standard) or degrees Rankine (°R) instead of °C or °F. Table 1.3-1 shows the equivalences of the four temperature scales.

TABLE 1.3-1. *Temperature Scales and Equivalents*

	Centigrade	Fahrenheit	Kelvin	Rankine	Celsius
Boiling water	100°C	212°F	373.15 K	671.67°R	100°C
Melting ice	0°C	32°F	273.15 K	491.67°R	0°C
Absolute zero	−273.15°C	−459.67°F	0 K	0°R	−273.15°C

The difference between the boiling point of water and melting point of ice at 1 atm is 100°C or 180°F. Thus, a 1.8°F change is equal to a 1°C change. Usually, the value of −273.15°C is rounded to −273.2°C and −459.67°F to −460°F. The following equations can be used to convert from one scale to another:

$$°F = 32 + 1.8(°C) \qquad (1.3\text{-}1)$$

$$°C = \frac{1}{1.8}(°F - 32) \qquad (1.3\text{-}2)$$

$$°R = °F + 459.67 \qquad (1.3\text{-}3)$$

$$K = °C + 273.15 \qquad (1.3\text{-}4)$$

1.3B Mole Units, and Weight or Mass Units

There are many methods used to express compositions in gases, liquids, and solids. One of the most useful is molar units, since chemical reactions and gas laws are simpler to express in terms of molar units. A mole (mol) of a pure substance is defined as the amount of that substance whose mass is numerically equal to its molecular weight. Hence, 1 kg mol of methane CH_4 contains 16.04 kg. Also, 1.0 lb mol contains 16.04 lb_m.

The mole fraction of a particular substance is simply the moles of this substance divided by the total number of moles. In like manner, the weight or mass fraction is the mass of the substance divided by the total mass. These two compositions, which hold for gases, liquids, and solids, can be expressed as follows for component A in a mixture:

$$x_A \text{ (mole fraction of } A) = \frac{\text{moles of } A}{\text{total moles}} \qquad (1.3\text{-}5)$$

$$w_A \text{ (mass or wt fraction of } A) = \frac{\text{mass } A}{\text{total mass}} \qquad (1.3\text{-}6)$$

EXAMPLE 1.3-1. Mole and Mass or Weight Fraction of a Solution

A container holds 50 g of water (B) and 50 g of NaOH (A). Calculate the weight fraction and mole fraction of NaOH. Also, calculate the lb_m of NaOH (A) and H_2O (B).

Solution: Taking as a basis for calculation 50 + 50 or 100 g of solution, the following data are calculated:

Component	G	Wt Fraction	Mol Wt	G Moles	Mole Fraction
H_2O (B)	50.0	$\frac{50}{100} = 0.500$	18.02	$\frac{50.0}{18.02} = 2.78$	$\frac{2.78}{4.03} = 0.690$
NaOH (A)	50.0	$\frac{50}{100} = 0.500$	40.0	$\frac{50.0}{40.0} = 1.25$	$\frac{1.25}{4.03} = 0.310$
Total	100.0	1.000		4.03	1.000

Hence, $x_A = 0.310$ and $x_B = 0.690$ and $x_A + x_B = 0.310 + 0.690 = 1.00$. Also, $w_A + w_B = 0.500 + 0.500 = 1.00$. To calculate the lb_m of each component, Appendix A.1 gives the conversion factor of 453.6 g per 1 lb_m. Using this,

$$\text{lb mass of } A = \frac{50 \text{ g } A}{453.6 \text{ g } A/\text{lb}_m A} = 0.1102 \text{ lb}_m A$$

Note that the g of A in the numerator cancels the g of A in the denominator, leaving lb_m of A in the numerator. The reader is cautioned to put all units down in an equation and cancel those appearing in the numerator and denominator. In a similar manner we obtain $0.1102 \text{ lb}_m B$ ($0.0500 \text{ kg } B$).

The analyses of solids and liquids are usually given as weight or mass fraction or weight percent, and gases as mole fraction or percent. Unless otherwise stated, analyses of solids and liquids will be assumed to be weight (mass) fraction or percent, and of gases to be mole fraction or percent.

1.3C   Concentration Units for Liquids

In general, when one liquid is mixed with another miscible liquid, the volumes are not additive. Hence, compositions of liquids are usually not expressed as volume percent of a component but as weight or mole percent. Another convenient way to express concentrations of components in a solution is *molarity,* which is defined as g mol of a component per liter of solution. Other methods used are kg/m^3, g/liter, g/cm^3, lb mol/cu ft, lb_m/cu ft, and lb_m/gallon. All these concentrations depend on temperature, so the temperature must be specified.

The most common method of expressing total concentration per unit volume is density, kg/m^3, g/cm^3, or lb_m/ft^3. For example, the density of water at 277.2 K ($4°C$) is 1000 kg/m^3, or $62.43 \text{ lb}_m/\text{ft}^3$. Sometimes the density of a solution is expressed as *specific gravity,* which is defined as the density of the solution at its given temperature divided by the density of a reference substance at its temperature. If the reference substance is water at 277.2 K, the specific gravity and density of the substance are numerically equal.

1.4   GAS LAWS AND VAPOR PRESSURE

1.4A   Pressure

There are numerous ways of expressing the pressure exerted by a fluid or system. An *absolute pressure* of 1.00 atm is equivalent to 760 mm Hg at $0°C$, 29.921 in. Hg, 0.760 m Hg, 14.696 lb force per square inch (psia), or 33.90 ft of water at $4°C$. *Gage pressure* is the pressure above the absolute pressure. Hence, a pressure of 21.5 lb per square inch gage (*psig*) is $21.5 + 14.7$ (rounded off), or 36.2 psia. In SI units, 1 psia $= 6.89476 \times 10^3$ pascal (Pa) $= 6.89476 \times 10^3$ newtons/m^2. Also, 1 atm $= 1.01325 \times 10^5$ Pa.

In some cases, particularly in evaporation, one may express the pressure as inches of mercury vacuum. This means the pressure as inches of mercury measured "below" the absolute barometric pressure. For example, a reading of 25.4 in. Hg vacuum is $29.92 - 25.4$, or 4.52 in. Hg absolute pressure. Pressure conversion units are given in Appendix A.1.

1.4B   Ideal Gas Law

An *ideal gas* is defined as one that obeys simple laws. Also, in an ideal gas the gas molecules are considered as rigid spheres which themselves occupy no volume and do not exert forces on one another. No real gases obey these laws exactly, but at ordinary temperatures and pressures of not more than several atmospheres, the ideal laws give answers within a few percent or less of the actual answers. Hence, these laws are sufficiently accurate for engineering calculations.

The *ideal gas law* of Boyle states that the volume of a gas is directly proportional to the absolute temperature and inversely proportional to the absolute pressure. This is expressed as

$$pV = nRT \tag{1.4-1}$$

where p is the absolute pressure in N/m^2, V the volume of the gas in m^3, n the kg mol of the gas, T the absolute temperature in K, and R the gas law constant of 8314.3 kg · m^2/kg mol · s^2 · K. When the volume is in ft^3, n in lb moles, and T in °R, R has a value of 0.7302 ft^3 · atm/lb mol · °R. For cgs units (see Appendix A.1), $V = cm^3$, $T = K$, $R = 82.057$ cm^3 · atm/g mol · K, and $n = $ g mol.

In order that amounts of various gases may be compared, *standard conditions of temperature and pressure* (abbreviated STP or SC) are arbitrarily defined as 101.325 kPa (1.0 atm) abs and 273.15 K (0°C). Under these conditions the volumes are as follows:

$$\text{volume of 1.0 kg mol (SC)} = 22.414 \text{ m}^3$$

$$\text{volume of 1.0 g mol (SC)} = 22.414 \text{ L (liter)}$$

$$= 22\ 414 \text{ cm}^3$$

$$\text{volume of 1.0 lb mol (SC)} = 359.05 \text{ ft}^3$$

EXAMPLE 1.4-1. Gas-Law Constant

Calculate the value of the gas-law constant R when the pressure is in psia, moles in lb mol, volume in ft^3, and temperature in °R. Repeat for SI units.

Solution: At standard conditions, $p = 14.7$ psia, $V = 359$ ft^3, and $T = 460 + 32 = 492$°R (273.15 K). Substituting into Eq. (1.4-1) for $n = 1.0$ lb mol and solving for R,

$$R = \frac{pV}{nT} = \frac{(14.7 \text{ psia})(359 \text{ ft}^3)}{(1.0 \text{ lb mol})(492°R)} = 10.73 \frac{\text{ft}^3 \cdot \text{psia}}{\text{lb mol} \cdot °R}$$

$$R = \frac{pV}{nT} = \frac{(1.01325 \times 10^5 \text{ Pa})(22.414 \text{ m}^3)}{(1.0 \text{ kg mol})(273.15 \text{ K})} = 8314 \frac{\text{m}^3 \cdot \text{Pa}}{\text{kg mol} \cdot \text{K}}$$

A useful relation can be obtained from Eq. (1.4-1) for n moles of gas at conditions p_1, V_1, T_1, and also at conditions p_2, V_2, T_2. Substituting into Eq. (1.4-1),

$$p_1V_1 = nRT_1$$

$$p_2V_2 = nRT_2$$

Combining gives

$$\frac{p_1V_1}{p_2V_2} = \frac{T_1}{T_2} \tag{1.4-2}$$

1.4C Ideal Gas Mixtures

Dalton's law for mixtures of ideal gases states that the total pressure of a gas mixture is equal to the sum of the individual partial pressures:

$$P = p_A + p_B + p_C + \cdots \qquad (1.4\text{-}3)$$

where P is total pressure and $p_A, p_B, p_C, \ldots$ are the partial pressures of the components A, B, $C, \ldots$ in the mixture.

Since the number of moles of a component is proportional to its partial pressure, the mole fraction of a component is

$$x_A = \frac{p_A}{P} = \frac{p_A}{p_A + p_B + p_C + \cdots} \qquad (1.4\text{-}4)$$

The volume fraction is equal to the mole fraction. Gas mixtures are almost always represented in terms of mole fractions and not weight fractions. For engineering purposes, Dalton's law is sufficiently accurate to use for actual mixtures at total pressures of a few atmospheres or less.

EXAMPLE 1.4-2. *Composition of a Gas Mixture*
A gas mixture contains the following components and partial pressures: CO_2, 75 mm Hg; CO, 50 mm Hg; N_2, 595 mm Hg; O_2, 26 mm Hg. Calculate the total pressure and the composition in mole fraction.

Solution: Substituting into Eq. (1.4-3),

$$P = p_A + p_B + p_C + p_D = 75 + 50 + 595 + 26 = 746 \text{ mm Hg}$$

The mole fraction of CO_2 is obtained by using Eq. (1.4-4).

$$x_A(CO_2) = \frac{p_A}{P} = \frac{75}{746} = 0.101$$

In like manner, the mole fractions of CO, N_2, and O_2 are calculated as 0.067, 0.797, and 0.035, respectively.

1.4D Vapor Pressure and Boiling Point of Liquids

When a liquid is placed in a sealed container, molecules of liquid will evaporate into the space above the liquid and fill it completely. After a time, equilibrium is reached. This vapor will exert a pressure just like a gas and we call this pressure the *vapor pressure* of the liquid. The value of the vapor pressure is independent of the amount of liquid in the container as long as some is present.

If an inert gas such as air is also present in the vapor space, it will have very little effect on the vapor pressure. In general, the effect of total pressure on vapor pressure can be considered as negligible for pressures of a few atmospheres or less.

The vapor pressure of a liquid increases markedly with temperature. For example, from Appendix A.2 for water, the vapor pressure at 50°C is 12.333 kPa (92.51 mm Hg). At 100°C the vapor pressure has increased greatly to 101.325 kPa (760 mm Hg).

The *boiling point* of a liquid is defined as the temperature at which the vapor pressure of a liquid equals the total pressure. Hence, if the atmospheric total pressure is 760 mm Hg, water will boil at 100°C. On top of a high mountain, where the total pressure is considerably less, water will boil at temperatures below 100°C.

A plot of vapor pressure P_A of a liquid versus temperature does not yield a straight line but a curve. However, for moderate temperature ranges, a plot of log P_A versus $1/T$ is a reasonably straight line, as follows:

$$\log P_A = m\left(\frac{1}{T}\right) + b \tag{1.4-5}$$

where m is the slope, b is a constant for the liquid A, and T is the temperature in K.

1.5 CONSERVATION OF MASS AND MATERIAL BALANCES

1.5A Conservation of Mass

One of the basic laws of physical science is the *law of conservation of mass*. This law, stated simply, says that mass cannot be created or destroyed (excluding, of course, nuclear or atomic reactions). Hence, the total mass (or weight) of all materials entering any process must equal the total mass of all materials leaving plus the mass of any materials accumulating or left in the process:

$$\text{input} = \text{output} + \text{accumulation} \tag{1.5-1}$$

In the majority of cases there will be no accumulation of materials in a process, and then the input will simply equal the output. Stated in other words, "what goes in must come out." We call this type of process a *steady-state process:*

$$\text{input} = \text{output (steady state)} \tag{1.5-2}$$

1.5B Simple Material Balances

In this section we do simple material (weight or mass) balances in various processes at steady state with no chemical reaction occurring. We can use units of kg, lb_m, lb mol, g, kg mol, and so on, in our balances. The reader is cautioned to be consistent and not to mix several units in a balance. When chemical reactions occur in the balances (as discussed in Section 1.5D), one should use kg mol units, since chemical equations relate moles reacting. In Section 2.6, overall mass balances will be covered in more detail and in Section 3.6, differential mass balances.

To solve a material-balance problem, it is advisable to proceed by a series of definite steps, as listed below:

1. *Sketch a simple diagram of the process.* This can be a simple box diagram showing each stream entering by an arrow pointing in and each stream leaving by an arrow pointing out. Include on each arrow the compositions, amounts, temperatures, and so on, of that stream. All pertinent data should be on this diagram.
2. *Write the chemical equations involved (if any).*
3. *Select a basis for calculation.* In most cases the problem is concerned with a specific amount of one of the streams in the process, which is selected as the basis.

4. *Make a material balance.* The arrows into the process will be input items and the arrows going out output items. The balance can be a total material balance in Eq. (1.5-2) or a balance on each component present (if no chemical reaction occurs).

Typical processes that do not undergo chemical reactions are drying, evaporation, dilution of solutions, distillation, extraction, and so on. These can be solved by setting up material balances containing unknowns and solving these equations for the unknowns.

EXAMPLE 1.5-1. Concentration of Orange Juice

In the concentration of orange juice, a fresh extracted and strained juice containing 7.08 wt % solids is fed to a vacuum evaporator. In the evaporator, water is removed and the solids content increased to 58 wt % solids. For 1000 kg/h entering, calculate the amounts of the outlet streams of concentrated juice and water.

Solution: Following the four steps outlined, we make a process flow diagram (step 1) in Fig. 1.5-1. Note that the letter W represents the unknown amount of water and C the amount of concentrated juice. No chemical reactions are given (step 2). Basis: 1000 kg/h entering juice (step 3).

To make the material balances (step 4), a total material balance will be made using Eq. (1.5-2):

$$1000 = W + C \tag{1.5-3}$$

This gives one equation and two unknowns. Hence, a component balance on solids will be made:

$$1000\left(\frac{7.08}{100}\right) = W(0) + C\left(\frac{58}{100}\right) \tag{1.5-4}$$

To solve these two equations, we solve Eq. (1.5-4) first for C since W drops out. We get $C = 122.1$ kg/h concentrated juice.

Substituting the value of C into Eq. (1.5-3),

$$1000 = W + 122.1$$

and we obtain $W = 877.9$ kg/h water.

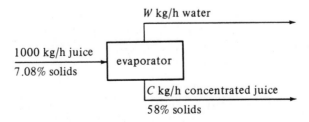

FIGURE 1.5-1. *Process flow diagram for Example 1.5-1.*

As a check on our calculations, we can write a balance on the water component:

$$1000\left(\frac{100 - 7.08}{100}\right) = 877.9 + 122.1\left(\frac{100 - 58}{100}\right) \qquad \textbf{(1.5-5)}$$

Solving,

$$929.2 = 877.9 + 51.3 = 929.2$$

In Example 1.5-1 only one unit or separate process was involved. Often, a number of processes in series are involved. Then we have a choice of making a separate balance over each separate process and/or a balance around the complete overall process.

1.5C   Material Balances and Recycle

Processes that have a recycle or feedback of part of the product into the entering feed are sometimes encountered. For example, in a sewage treatment plant, part of the activated sludge from a sedimentation tank is recycled back to the aeration tank where the liquid is treated. In some food-drying operations, the humidity of the entering air is controlled by re-circulating part of the hot, wet air that leaves the dryer. In chemical reactions, the material that did not react in the reactor can be separated from the final product and fed back to the reactor.

EXAMPLE 1.5-2.   Crystallization of KNO_3 and Recycle

In a process producing KNO_3 salt, 1000 kg/h of a feed solution containing 20 wt % KNO_3 is fed to an evaporator, which evaporates some water at 422 K to produce a 50 wt % KNO_3 solution. This is then fed to a crystallizer at 311 K, where crystals containing 96 wt % KNO_3 are removed. The saturated solution containing 37.5 wt % KNO_3 is recycled to the evaporator. Calculate the amount of recycle stream R in kg/h and the product stream of crystals P in kg/h.

Solution: Figure 1.5-2 gives the process flow diagram. As a basis we shall use 1000 kg/h of fresh feed. No chemical reactions are occurring. We can make an overall balance on the entire process for KNO_3 and solve for P directly:

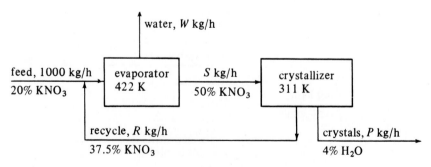

FIGURE 1.5-2.   *Process flow diagram for Example 1.5-2.*

$$1000(0.20) = W(0) + P(0.96) \tag{1.5-6}$$

$$P = 208.3 \text{ kg crystals/h}$$

To calculate the recycle stream, we can make a balance around the evaporator or the crystallizer. Using a balance on the crystallizer, since it now includes only two unknowns, S and R, we get for a total balance,

$$S = R + 208.3 \tag{1.5-7}$$

For a KNO_3 balance on the crystallizer,

$$S(0.50) = R(0.375) + 208.3(0.96) \tag{1.5-8}$$

Substituting S from Eq. (1.5-7) into Eq. (1.5-8) and solving, $R = 766.6$ kg recycle/h and $S = 974.9$ kg/h.

1.5D Material Balances and Chemical Reaction

In many cases the materials entering a process undergo chemical reactions in the process, so that the materials leaving are different from those entering. In these cases it is usually convenient to make a molar and not a weight balance on an individual component, such as kg mol H_2 or kg atom H, kg mol CO_3^- ion, kg mol $CaCO_3$, kg atom Na^+, kg mol N_2, and so on. For example, in the combustion of CH_4 with air, balances can be made on kg mol of H_2, C, O_2, or N_2.

EXAMPLE 1.5-3. *Combustion of Fuel Gas*
A fuel gas containing 3.1 mol % H_2, 27.2% CO, 5.6% CO_2, 0.5% O_2, and 63.6% N_2 is burned with 20% excess air (i.e., the air over and above that necessary for complete combustion to CO_2 and H_2O). The combustion of CO is only 98% complete. For 100 kg mol of fuel gas, calculate the moles of each component in the exit flue gas.

Solution: First, the process flow diagram is drawn (Fig. 1.5-3). On the diagram the components in the flue gas are shown. Let A be moles of air and F be moles of flue gas. Next the chemical reactions are given:

$$CO + \tfrac{1}{2}O_2 \rightarrow CO_2 \tag{1.5-9}$$

$$H_2 + \tfrac{1}{2}O_2 \rightarrow H_2O \tag{1.5-10}$$

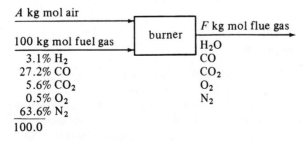

FIGURE 1.5-3. *Process flow diagram for Example 1.5-3.*

An accounting of the total moles of O_2 in the fuel gas is as follows:

$$\text{mol } O_2 \text{ in fuel gas} = (\tfrac{1}{2})27.2(CO) + 5.6(CO_2) + 0.5(O_2) = 19.7 \text{ mol } O_2$$

For all the H_2 to be completely burned to H_2O, we need, from Eq. (1.5-10), $\tfrac{1}{2}$ mol O_2 for 1 mol H_2, or $3.1(\tfrac{1}{2}) = 1.55$ total mol O_2. For completely burning the CO from Eq. (1.5-9), we need $27.2(\tfrac{1}{2}) = 13.6$ mol O_2. Hence, the amount of O_2 we must add is, theoretically, as follows:

$$\text{mol } O_2 \text{ theoretically needed} = 1.55 + 13.6 - 0.5 \text{ (in fuel gas)}$$

$$= 14.65 \text{ mol } O_2$$

For a 20% excess, we add 1.2(14.65), or 17.58 mol O_2. Since air contains 79 mol % N_2, the amount of N_2 added is (79/21)(17.58), or 66.1 mol N_2.

To calculate the moles in the final flue gas, all the H_2 gives H_2O, or 3.1 mol H_2O. For CO, 2.0% does not react. Hence, 0.02(27.2), or 0.54 mol CO will be unburned.

A total carbon balance is as follows: inlet moles C = 27.2 + 5.6 = 32.8 mol C. In the outlet flue gas, 0.54 mol will be as CO and the remainder of 32.8 − 0.54, or 32.26 mol as CO_2.

For calculating the outlet mol O_2, we make an overall O_2 balance:

$$O_2 \text{ in} = 19.7 \text{ (in fuel gas)} + 17.58 \text{ (in air)} = 37.28 \text{ mol } O_2$$

$$O_2 \text{ out} = (3.1/2) \text{ (in } H_2O) + (0.54/2) \text{ (in CO)} + 32.26 \text{ (in } CO_2) + \text{free } O_2$$

Equating inlet O_2 to outlet, the free remaining $O_2 = 3.2$ mol O_2. For the N_2 balance, the outlet = 63.6 (in fuel gas) + 66.1 (in air), or 129.70 mol N_2. The outlet flue gas contains 3.10 mol H_2O, 0.54 mol CO, 32.26 mol CO_2, 3.20 mol O_2, and 129.7 mol N_2.

In chemical reactions with several reactants, the limiting reactant component is defined as that compound which is present in an amount less than the amount necessary for it to react stoichiometrically with the other reactants. Then the percent completion of a reaction is the amount of this limiting reactant actually converted, divided by the amount originally present, times 100.

1.6 ENERGY AND HEAT UNITS

1.6A Joule, Calorie, and Btu

In a manner similar to that used in making material balances on chemical and biological processes, we can also make energy balances on a process. Often a large portion of the energy entering or leaving a system is in the form of heat. Before such energy or heat balances are made, we must understand the various types of energy and heat units.

In the SI system, energy is given in joules (J) or kilojoules (kJ). Energy is also expressed in btu (British thermal units) or cal (calories). The g calorie (abbreviated cal) is defined as the amount of heat needed to heat 1.0 g water 1.0°C (from 14.5°C to 15.5°C). Also, 1 kcal (kilocalorie) = 1000 cal. The btu is defined as the amount of heat needed to raise 1.0 lb water 1°F. Hence, from Appendix A.1,

$$1 \text{ btu} = 252.16 \text{ cal} = 1.05506 \text{ kJ} \qquad \textbf{(1.6-1)}$$

1.6B Heat Capacity

The *heat capacity* of a substance is defined as the amount of heat necessary to increase the temperature by 1 degree. It can be expressed for 1 g, 1 lb, 1 g mol, 1 kg mol, or 1 lb mol of the substance. For example, a heat capacity is expressed in SI units as J/kg mol · K; in other units as cal/g · °C, cal/g mol · °C, kcal/kg mol · °C, btu/lb$_m$ · °F, or btu/lb mol · °F.

It can be shown that the actual numerical value of a heat capacity is the same in mass units or in molar units. That is,

$$1.0 \text{ cal/g} \cdot °C = 1.0 \text{ btu/lb}_m \cdot °F \qquad \textbf{(1.6-2)}$$

$$1.0 \text{ cal/g mol} \cdot °C = 1.0 \text{ btu/lb mol} \cdot °F \qquad \textbf{(1.6-3)}$$

For example, to prove this, suppose that a substance has a heat capacity of 0.8 btu/lb$_m$ · °F. The conversion is made using 1.8°F for 1°C or 1 K, 252.16 cal for 1 btu, and 453.6 g for 1 lb$_m$, as follows:

$$\text{heat capacity} \left(\frac{\text{cal}}{\text{g} \cdot °C} \right) = \left(0.8 \, \frac{\text{btu}}{\text{lb}_m \cdot °F} \right) \left(252.16 \, \frac{\text{cal}}{\text{btu}} \right) \left(\frac{1}{453.6 \text{ g/lb}_m} \right) \left(1.8 \, \frac{°F}{°C} \right)$$

$$= 0.8 \, \frac{\text{cal}}{\text{g} \cdot °C}$$

The heat capacities of gases (sometimes called *specific heat*) at constant pressure c_p are functions of temperature and for engineering purposes can be assumed to be independent of pressure up to several atmospheres. In most process engineering calculations, one is usually interested in the amount of heat needed to heat a gas from one temperature t_1 to another at t_2. Since the c_p varies with temperature, an integration must be performed or a suitable mean c_{pm} used. These mean values for gases have been obtained for T_1 of 298 K or 25°C (77°F) and various T_2 values, and are tabulated in Table 1.6-1 at 101.325 kPa pressure or less as c_{pm} in kJ/kg mol · K at various values of T_2 in K or °C.

EXAMPLE 1.6-1. Heating of N$_2$ Gas
The gas N$_2$ at 1 atm pressure absolute is being heated in a heat exchanger. Calculate the amount of heat needed in J to heat 3.0 g mol N$_2$ in the following temperature ranges:
(a) 298–673 K (25–400°C)
(b) 298–1123 K (25–850°C)
(c) 673–1123 K (400–850°C)

Solution: For case (a), Table 1.6-1 gives c_{pm} values at 1 atm pressure or less which can be used up to several atm pressures. For N$_2$ at 673 K, c_{pm} = 29.68 kJ/kg mol · K or 29.68 J/g mol · K. This is the mean heat capacity for the range 298–673 K:

$$\text{heat required} = M \text{ g mol} \left(c_{pm} \, \frac{J}{\text{g mol} \cdot K} \right) (T_2 - T_1)K \qquad \textbf{(1.6-4)}$$

TABLE 1.6-1. *Mean Molar Heat Capacities of Gases Between 298 and TK (25 and T°C) at 101.325 kPa or Less (SI Units: c_p = kJ/kg mol · K)*

T(K)	T(°C)	H_2	N_2	CO	Air	O_2	H_2O	CO_2	CH_4	SO_2
298	25	28.86	29.14	29.16	29.19	29.38	33.59	37.20	35.8	39.9
373	100	28.99	29.19	29.24	29.29	29.66	33.85	38.73	37.6	41.2
473	200	29.13	29.29	29.38	29.40	30.07	34.24	40.62	40.3	42.9
573	300	29.18	29.46	29.60	29.61	30.53	34.39	42.32	43.1	44.5
673	400	29.23	29.68	29.88	29.94	31.01	35.21	43.80	45.9	45.8
773	500	29.29	29.97	30.19	30.25	31.46	35.75	45.12	48.8	47.0
873	600	29.35	30.27	30.52	30.56	31.89	36.33	46.28	51.4	47.9
973	700	29.44	30.56	30.84	30.87	32.26	36.91	47.32	54.0	48.8
1073	800	29.56	30.85	31.16	31.18	32.62	37.53	48.27	56.4	49.6
1173	900	29.63	31.16	31.49	31.48	32.97	38.14	49.15	58.8	50.3
1273	1000	29.84	31.43	31.77	31.79	33.25	38.71	49.91	61.0	50.9
1473	1200	30.18	31.97	32.30	32.32	33.78	39.88	51.29	64.9	51.9
1673	1400	30.51	32.40	32.73	32.76	34.19	40.90	52.34		

Mean Molar Heat Capacities of Gases Between 25 and T°C at 1 atm Pressure or Less (English Units: c_p = btu/lb mol · °F)

T(°C)	H_2	N_2	CO	Air	O_2	NO	H_2O	CO_2	HCl	Cl_2	CH_4	SO_2	C_2H_4	SO_3	C_2H_6
25	6.894	6.961	6.965	6.972	7.017	7.134	8.024	8.884	6.96	8.12	8.55	9.54	10.45	12.11	12.63
100	6.924	6.972	6.983	6.996	7.083	7.144	8.084	9.251	6.97	8.24	8.98	9.85	11.35	12.84	13.76
200	6.957	6.996	7.017	7.021	7.181	7.224	8.177	9.701	6.98	8.37	9.62	10.25	12.53	13.74	15.27
300	6.970	7.036	7.070	7.073	7.293	7.252	8.215	10.108	7.00	8.48	10.29	10.62	13.65	14.54	16.72
400	6.982	7.089	7.136	7.152	7.406	7.301	8.409	10.462	7.02	8.55	10.97	10.94	14.67	15.22	18.11
500	6.995	7.159	7.210	7.225	7.515	7.389	8.539	10.776	7.06	8.61	11.65	11.22	15.60	15.82	19.39
600	7.011	7.229	7.289	7.299	7.616	7.470	8.678	11.053	7.10	8.66	12.27	11.45	16.45	16.33	20.58
700	7.032	7.298	7.365	7.374	7.706	7.549	8.816	11.303	7.15	8.70	12.90	11.66	17.22	16.77	21.68
800	7.060	7.369	7.443	7.447	7.792	7.630	8.963	11.53	7.21	8.73	13.48	11.84	17.95	17.17	22.72
900	7.076	7.443	7.521	7.520	7.874	7.708	9.109	11.74	7.27	8.77	14.04	12.01	18.63	17.52	23.69
1000	7.128	7.507	7.587	7.593	7.941	7.773	9.246	11.92	7.33	8.80	14.56	12.15	19.23	17.86	24.56
1100	7.169	7.574	7.653	7.660	8.009	7.839	9.389	12.10	7.39	8.82	15.04	12.28	19.81	18.17	25.40
1200	7.209	7.635	7.714	7.719	8.068	7.898	9.524	12.25	7.45	8.94	15.49	12.39	20.33	18.44	26.15
1300	7.252	7.692	7.772	7.778	8.123	7.952	9.66	12.39							
1400	7.288	7.738	7.818	7.824	8.166	7.994	9.77	12.50							
1500	7.326	7.786	7.866	7.873	8.203	8.039	9.89	12.69							
1600	7.386	7.844	7.922	7.929	8.269	8.092	9.95	12.75							
1700	7.421	7.879	7.958	7.965	8.305	8.124	10.13	12.70							
1800	7.467	7.924	8.001	8.010	8.349	8.164	10.24	12.94							
1900	7.505	7.957	8.033	8.043	8.383	8.192	10.34	13.01							
2000	7.548	7.994	8.069	8.081	8.423	8.225	10.43	13.10							
2100	7.588	8.028	8.101	8.115	8.460	8.255	10.52	13.17							
2200	7.624	8.054	8.127	8.144	8.491	8.277	10.61	13.24							

Source: O. A. Hougen, K. W. Watson, and R. A. Ragatz, *Chemical Process Principles,* Part I, 2nd ed. New York: John Wiley & Sons, Inc., 1954. With permission.

Substituting the known values,

$$\text{heat required} = (3.0)(29.68)(673 - 298) = 33\,390 \text{ J}$$

For case (b), the c_{pm} at 1123 K (obtained by linear interpolation between 1073 and 1173 K) is 31.00 J/g mol · K:

$$\text{heat required} = (3.0)(31.00)(1123 - 298) = 76\,725 \text{ J}$$

For case (c), there is no mean heat capacity for the interval 673–1123 K. However, we can use the heat required to heat the gas from 298 to 673 K in case (a) and subtract it from case (b), which includes the heat to go from 298 to 673 K plus 673 to 1123 K:

$$\text{heat required (673–1123 K)} = \text{heat required (298–1123 K)}$$

$$- \text{heat required (298–673)} \qquad \textbf{(1.6-5)}$$

Substituting the proper values into Eq. (1.6-5),

$$\text{heat required} = 76\,725 - 33\,390 = 43\,335 \text{ J}$$

On heating a gas mixture, the total heat required is determined by first calculating the heat required for each individual component and then adding the results to obtain the total.

The heat capacities of solids and liquids are also functions of temperature and independent of pressure. Data are given in Appendix A.2, Physical Properties of Water; A.3, Physical Properties of Inorganic and Organic Compounds; and A.4, Physical Properties of Foods and Biological Materials. More data are available in (P1).

EXAMPLE 1.6-2. Heating of Milk

Rich cows' milk (4536 kg/h) at 4.4°C is being heated in a heat exchanger to 54.4°C by hot water. How much heat is needed?

Solution: From Appendix A.4, the average heat capacity of rich cows' milk is 3.85 kJ/kg · K. Temperature rise $\Delta T = (54.4 - 4.4)°C = 50$ K.

$$\text{heat required} = (4536 \text{ kg/h})(3.85 \text{ kJ/kg · K})(1/3600 \text{ h/s})(50 \text{ K}) = 242.5 \text{ kW}$$

The enthalpy, H, of a substance in J/kg represents the sum of the internal energy plus the pressure–volume term. For no reaction and a constant-pressure process with a change in temperature, the heat change as computed from Eq. (1.6-4) is the difference in enthalpy, ΔH, of the substance relative to a given temperature or base point. In other units, $H = \text{btu/lb}_m$ or cal/g.

1.6C Latent Heat and Steam Tables

Whenever a substance undergoes a change of phase, relatively large amounts of heat change are involved at a constant temperature. For example, ice at 0°C and 1 atm pressure can absorb 6013.4 kJ/kg mol. This enthalpy change is called the *latent heat of fusion*. Data for other compounds are available in various handbooks (P1, W1).

When a liquid phase vaporizes to a vapor phase under its vapor pressure at constant temperature, an amount of heat called the *latent heat of vaporization* must be added. Tabulations of latent heats of vaporization are given in various handbooks. For water at 25°C and a pressure of 23.75 mm Hg, the latent heat is 44 020 kJ/kg mol, and at 25°C and 760 mm Hg, 44 045 kJ/kg mol. Hence, the effect of pressure can be neglected in engineering calculations. However, there is a large effect of temperature on the latent heat of water. Also, the effect of pressure on the heat capacity of liquid water is small and can be neglected.

Since water is a very common chemical, the thermodynamic properties of it have been compiled in steam tables and are given in Appendix A.2 in SI and in English units.

EXAMPLE 1.6-3. *Use of Steam Tables*

Find the enthalpy change (i.e., how much heat must be added) for each of the following cases using SI and English units:

    (a) Heating 1 kg (lb_m) water from 21.11°C (70°F) to 60°C (140°F) at 101.325 kPa (1 atm) pressure.

    (b) Heating 1 kg (lb_m) water from 21.11°C (70°F) to 115.6°C (240°F) and vaporizing at 172.2 kPa (24.97 psia).

    (c) Vaporizing 1 kg (lb_m) water at 115.6°C (240°F) and 172.2 kPa (24.97 psia).

Solution: For part (a), the effect of pressure on the enthalpy of liquid water is negligible. From Appendix A.2,

$$H \text{ at } 21.11°C = 88.60 \text{ kJ/kg} \quad \text{or} \quad \text{at } 70°F = 38.09 \text{ btu/lb}_m$$

$$H \text{ at } 60°C = 251.13 \text{ kJ/kg} \quad \text{or} \quad \text{at } 140°F = 107.96 \text{ btu/lb}_m$$

$$\text{change in } H = \Delta H = 251.13 - 88.60 = 162.53 \text{ kJ/kg}$$

$$= 107.96 - 38.09 = 69.87 \text{ btu/lb}_m$$

In part (b), the enthalpy at 115.6°C (240°F) and 172.2 kPa (24.97 psia) of the saturated vapor is 2699.9 kJ/kg or 1160.7 btu/lb_m.

$$\text{change in } H = \Delta H = 2699.9 - 88.60 = 2611.3 \text{ kJ/kg}$$

$$= 1160.7 - 38.09 = 1122.6 \text{ btu/lb}_m$$

The latent heat of water at 115.6°C (240°F) in part (c) is

$$2699.9 - 484.9 = 2215.0 \text{ kJ/kg}$$

$$1160.7 - 208.44 = 952.26 \text{ btu/lb}_m$$

1.6D Heat of Reaction

When chemical reactions occur, heat effects always accompany these reactions. This area where energy changes occur is often called *thermochemistry*. For example, when HCl is neutralized with NaOH, heat is given off and the reaction is exothermic. Heat is absorbed in an endothermic reaction. This heat of reaction is dependent on the chemical nature of each reacting material and product and on their physical states.

For purposes of organizing data, we define a standard heat of reaction ΔH^0 as the change in enthalpy when 1 kg mol reacts under a pressure of 101.325 kPa at a temperature of 298 K (25°C). For example, for the reaction

$$H_2(g) + \tfrac{1}{2}O_2(g) \rightarrow H_2O(l) \tag{1.6-6}$$

the ΔH^0 is -285.840×10^3 kJ/kg mol or -68.317 kcal/g mol. The reaction is exothermic and the value is negative since the reaction loses enthalpy. In this case, the H_2 gas reacts with the O_2 gas to give liquid water, all at 298 K (25°C).

Special names are given to ΔH^0 depending upon the type of reaction. When the product is formed from the elements, as in Eq. (1.6-6), we call the ΔH^0 the *heat of formation* of the product water, ΔH_f^0. For the combustion of CH_4 to form CO_2 and H_2O, we call it *heat of combustion, ΔH_c^0.* Data are given in Appendix A.3 for various values of ΔH_c^0.

EXAMPLE 1.6-4.   Combustion of Carbon

A total of 10.0 g mol of carbon graphite is burned in a calorimeter held at 298 K and 1 atm. The combustion is incomplete, and 90% of the C goes to CO_2 and 10% to CO. What is the total enthalpy change in kJ and kcal?

Solution: From Appendix A.3 the ΔH_c^0 for carbon going to CO_2 is -393.513×10^3 kJ/kg mol or -94.0518 kcal/g mol, and for carbon going to CO it is -110.523×10^3 kJ/kg mol or -26.4157 kcal/g mol. Since 9 mol CO_2 and 1 mol CO are formed,

$$\text{total } \Delta H = 9(-393.513) + 1(-110.523) = -3652 \text{ kJ}$$

$$= 9(-94.0518) + 1(-26.4157) = -872.9 \text{ kcal}$$

If a table of heats of formation, ΔH_f^0, of compounds is available, the standard heat of the reaction, ΔH^0, can be calculated by

$$\Delta H^0 = \sum \Delta H_{f\,(products)}^0 - \sum \Delta H_{f\,(reactants)}^0 \tag{1.6-7}$$

In Appendix A.3, a short table of some values of ΔH_f is given. Other data are also available (H1, P1, S1).

EXAMPLE 1.6-5.   Reaction of Methane

For the following reaction of 1 kg mol of CH_4 at 101.32 kPa and 298 K,

$$CH_4(g) + H_2O(l) \rightarrow CO(g) + 3H_2(g)$$

calculate the standard heat of reaction ΔH^0 at 298 K in kJ.

Solution: From Appendix A.3, the following standard heats of formation are obtained at 298 K:

	ΔH_f^0 (kJ/kg mol)
$CH_4(g)$	-74.848×10^3
$H_2O(l)$	-285.840×10^3
$CO(g)$	-110.523×10^3
$H_2(g)$	0

Note that the ΔH_f^0 of all elements is, by definition, zero. Substituting into Eq. (1.6-7),

$$\Delta H^0 = [-110.523 \times 10^3 - 3(0)] - (-74.848 \times 10^3 - 285.840 \times 10^3)$$

$$= +250.165 \times 10^3 \text{ kJ/kg mol} \qquad \text{(endothermic)}$$

1.7 CONSERVATION OF ENERGY AND HEAT BALANCES

1.7A Conservation of Energy

In making material balances we used the law of conservation of mass, which states that the mass entering is equal to the mass leaving plus the mass left in the process. In a similar manner, we can state the *law of conservation of energy,* which says that all energy entering a process is equal to that leaving plus that left in the process. In this section elementary heat balances will be made. More elaborate energy balances will be considered in Sections 2.7 and 5.6.

Energy can appear in many forms. Some of the common forms are enthalpy, electrical energy, chemical energy (in terms of ΔH reaction), kinetic energy, potential energy, work, and heat inflow.

In many cases in process engineering, which often takes place at constant pressure, electrical energy, kinetic energy, potential energy, and work either are not present or can be neglected. Then only the enthalpy of the materials (at constant pressure), the standard chemical reaction energy (ΔH^0) at 25°C, and the heat added or removed must be taken into account in the energy balance. This is generally called a *heat balance.*

1.7B Heat Balances

In making a heat balance at steady state we use methods similar to those used in making a material balance. The energy or heat coming into a process in the inlet materials plus any net energy added to the process are equal to the energy leaving in the materials. Expressed mathematically,

$$\sum H_R + (-\Delta H_{298}^0) + q = \sum H_p \qquad \textbf{(1.7-1)}$$

where $\sum H_R$ is the sum of enthalpies of all materials entering the reaction process relative to the reference state for the standard heat of reaction at 298 K and 101.32 kPa. If the inlet temperature is above 298 K, this sum will be positive. $\Delta H_{298}^0 =$ standard heat of the reaction at 298 K and 101.32 kPa. The reaction contributes heat to the process, so the negative of ΔH_{298}^0 is taken to be positive input heat for an exothermic reaction. Also, $q =$ net energy or heat added to the system. If heat leaves the system, this item will be negative. $\sum H_p =$ sum of enthalpies of all leaving materials referred to the standard reference state at 298 K (25°C).

Note that if the materials coming into a process are below 298 K, $\sum H_R$ will be negative. Care must be taken not to confuse the signs of the items in Eq. (1.7-1). If no chemical reaction occurs, then simple heating, cooling, or phase change is occurring. Use of Eq. (1.7-1) will be illustrated by several examples. For convenience it is common practice to call the terms on the left-hand side of Eq. (1.7-1) input items, and those on the right, output items.

EXAMPLE 1.7-1. *Heating of Fermentation Medium*
A liquid fermentation medium at 30°C is pumped at a rate of 2000 kg/h through a heater, where it is heated to 70°C under pressure. The waste heat water used to

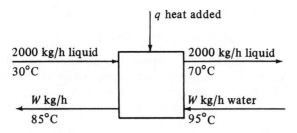

FIGURE 1.7-1. *Process flow diagram for Example 1.7-1.*

heat this medium enters at 95°C and leaves at 85°C. The average heat capacity of the fermentation medium is 4.06 kJ/kg·K, and that for water is 4.21 kJ/kg·K (Appendix A.2). The fermentation stream and the wastewater stream are separated by a metal surface through which heat is transferred and do not physically mix with each other. Make a complete heat balance on the system. Calculate the water flow and the amount of heat added to the fermentation medium assuming no heat losses. The process flow is given in Fig. 1.7-1.

Solution: It is convenient to use the standard reference state of 298 K (25°C) as the datum to calculate the various enthalpies. From Eq. (1.7-1) the input items are as follows:

Input items. ΣH_R of the enthalpies of the two streams relative to 298 K (25°C) (note that $\Delta t = 30 - 25°C = 5°C = 5$ K):

$$H(\text{liquid}) = (2000 \text{ kg/h})(4.06 \text{ kJ/kg} \cdot \text{K})(5 \text{ K})$$

$$= 4.060 \times 10^4 \text{ kJ/h}$$

$$H(\text{water}) = W(4.21)(95 - 25) = 2.947 \times 10^2 \, W \text{ kJ/h} \qquad (W = \text{kg/h})$$

$$(-\Delta H^0_{298}) = 0 \qquad \text{(since there is no chemical reaction)}$$

$$q = 0 \qquad \text{(there are no heat losses or additions)}$$

Output items. ΣH_P of the two streams relative to 298 K (25°C):

$$H(\text{liquid}) = 2000(4.06)(70 - 25) = 3.65 \times 10^5 \text{ kJ/h}$$

$$H(\text{water}) = W(4.21)(85 - 25) = 2.526 \times 10^2 \, W \text{ kJ/h}$$

Equating input to output in Eq. (1.7-1) and solving for W,

$$4.060 \times 10^4 + 2.947 \times 10^2 \, W = 3.654 \times 10^5 + 2.526 \times 10^2 \, W$$

$$W = 7720 \text{ kg/h water flow}$$

The amount of heat added to the fermentation medium is simply the difference of the outlet and inlet liquid enthalpies:

$$H(\text{outlet liquid}) - H(\text{inlet liquid}) = 3.654 \times 10^5 - 4.060 \times 10^4$$

$$= 3.248 \times 10^5 \text{ kJ/h (90.25 kW)}$$

Note in this example that since the heat capacities were assumed constant, a simpler balance could have been written as follows:

$$\text{heat gained by liquid} = \text{heat lost by water}$$

$$2000(4.06)(70 - 30) = W(4.21)(95 - 85)$$

Then, solving, $W = 7720$ kg/h. This simple balance works well when c_p is constant. However, when c_p varies with temperature and the material is a gas, c_{pm} values are only available between 298 K (25°C) and t K, and the simple method cannot be used without obtaining new c_{pm} values over different temperature ranges.

EXAMPLE 1.7-2   Heat and Material Balance in Combustion

The waste gas from a process of 1000 g mol/h of CO at 473 K is burned at 1 atm pressure in a furnace using air at 373 K. The combustion is complete and 90% excess air is used. The flue gas leaves the furnace at 1273 K. Calculate the heat removed in the furnace.

Solution: First, the process flow diagram is drawn in Fig. 1.7-2, and then a material balance is made:

$$CO(g) + \tfrac{1}{2}O_2(g) \rightarrow CO_2(g)$$

$$\Delta H^0_{298} = -282.989 \times 10^3 \text{ kJ/kg mol}$$

$$\text{(from Appendix A.3)}$$

$$\text{mol CO} = 1000 \text{ g mol/h} = \text{moles CO}_2$$

$$= 1.00 \text{ kg mol/h}$$

$$\text{mol O}_2 \text{ theoretically required} = \tfrac{1}{2}(1.00) = 0.500 \text{ kg mol/h}$$

$$\text{mol O}_2 \text{ actually added} = 0.500(1.9) = 0.950 \text{ kg mol/h}$$

$$\text{mol N}_2 \text{ added} = 0.950 \frac{0.79}{0.21} = 3.570 \text{ kg mol/h}$$

$$\text{air added} = 0.950 + 3.570 = 4.520 \text{ kg mol/h} = A$$

$$\text{O}_2 \text{ in outlet flue gas} = \text{added} - \text{used}$$

$$= 0.950 - 0.500 = 0.450 \text{ kg mol/h}$$

$$\text{CO}_2 \text{ in outlet flue gas} = 1.00 \text{ kg mol/h}$$

$$\text{N}_2 \text{ in outlet flue gas} = 3.570 \text{ kg mol/h}$$

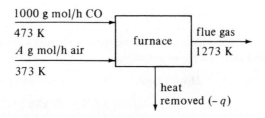

1000 g mol/h CO

473 K

A g mol/h air

373 K

furnace

flue gas

1273 K

heat removed $(-q)$

FIGURE 1.7-2.   *Process flow diagram for Example 1.7-2.*

For the heat balance relative to the standard state at 298 K, we follow Eq. (1.7-1).

Input items

$$H(CO) = 1.00(c_{pm})(473 - 298) = 1.00(29.38)(473 - 298) = 5142 \text{ kJ/h}$$

(The c_{pm} of CO of 29.38 kJ/kg mol $\cdot$ K between 298 and 473 K is obtained from Table 1.6-1.)

$$H(\text{air}) = 4.520(c_{pm})(373 - 298) = 4.520(29.29)(373 - 298) = 9929 \text{ kJ/h}$$

$$q = \text{heat added, kJ/h}$$

(This will give a negative value here, indicating that heat was removed.)

$$-\Delta H^0_{298} = -(-282.989 \times 10^3 \text{ kJ/kg mol})(1.00 \text{ kg mol/h}) = 282\,990 \text{ kJ/h}$$

Output items

$$H(CO_2) = 1.00(c_{pm})(1273 - 298) = 1.00(49.91)(1273 - 298) = 48\,660 \text{ kJ/h}$$

$$H(O_2) = 0.450(c_{pm})(1273 - 298) = 0.450(33.25)(1273 - 298) = 14\,590 \text{ kJ/h}$$

$$H(N_2) = 3.570(c_{pm})(1273 - 298) = 3.570(31.43)(1273 - 298) = 109\,400 \text{ kJ/h}$$

Equating input to output and solving for q,

$$5142 + 9929 + q + 282\,990 = 48\,660 + 14\,590 + 109\,400$$

$$q = -125\,411 \text{ kJ/h}$$

Hence, heat is removed: $-34\,837$ W.

Often when chemical reactions occur in the process and the heat capacities vary with temperature, the solution in a heat balance can be trial and error if the final temperature is the unknown.

EXAMPLE 1.7-3. Oxidation of Lactose

In many biochemical processes, lactose is used as a nutrient, which is oxidized as follows:

$$C_{12}H_{22}O_{11}(s) + 12O_2(g) \rightarrow 12CO_2(g) + 11H_2O(l)$$

The heat of combustion ΔH^0_c in Appendix A.3 at 25°C is -5648.8×10^3 J/g mol. Calculate the heat of complete oxidation (combustion) at 37°C, which is the temperature of many biochemical reactions. The c_{pm} of solid lactose is 1.20 J/g $\cdot$ K, and the molecular weight is 342.3 g mass/g mol.

Solution: This can be treated as an ordinary heat-balance problem. First, the process flow diagram is drawn in Fig. 1.7-3. Next, the datum temperature of 25°C is selected and the input and output enthalpies calculated. The temperature difference $\Delta t = (37 - 25)$°C = $(37 - 25)$ K.

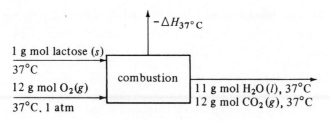

FIGURE 1.7-3. *Process flow diagram for Example 1.7-3.*

Input items

$$H(\text{lactose}) = (342.3 \text{ g})\left(c_{pm}\frac{J}{g \cdot K}\right)(37 - 25) \text{ K} = 342.3(1.20)(37 - 25)$$

$$= 4929 \text{ J}$$

$$H(\text{O}_2 \text{ gas}) = (12 \text{ g mol})\left(c_{pm}\frac{J}{g \text{ mol} \cdot K}\right)(37 - 25) \text{ K}$$

$$= 12(29.38)(37 - 25) = 4230 \text{ J}$$

(The c_{pm} of O_2 was obtained from Table 1.6-1.)

$$-\Delta H_{25}^{0} = -(-5648.8 \times 10^{3})$$

Output items

$$H(\text{H}_2\text{O liquid}) = 11(18.02 \text{ g})\left(c_{pm}\frac{J}{g \cdot K}\right)(37 - 25) \text{ K}$$

$$= 11(18.02)(4.18)(37 - 25) = 9943 \text{ J}$$

(The c_{pm} of liquid water was obtained from Appendix A.2.)

$$H(\text{CO}_2 \text{ gas}) = (12 \text{ g mol})\left(c_{pm}\frac{J}{g \text{ mol} \cdot K}\right)(37 - 25) \text{ K}$$

$$= 12(37.45)(37 - 25) = 5393 \text{ J}$$

(The c_{pm} of CO_2 is obtained from Table 1.6-1.)

$\Delta H_{37°C}$:
Setting input = output and solving,

$$4929 + 4230 + 5648.8 \times 10^{3} = 9943 + 5393 - \Delta H_{37°c}$$

$$\Delta H_{37°C} = -5642.6 \times 10^{3} \text{ J/g mol} = \Delta H_{310 \text{ K}}$$

1.8   NUMERICAL METHODS FOR INTEGRATION

1.8A   Introduction and Graphical Integration

Often the mathematical function $f(x)$ to be integrated is too complex and we are not able to integrate it analytically. Or in some cases the function is one that has been obtained from experimental data, and no mathematical equation is available to represent the data so that they can be integrated analytically. In these cases, we can use either numerical or graphical integration.

Integration of a function $f(x)$ between the limits $x = a$ to $x = b$ can be represented by

$$\int_{x=a}^{x=b} f(x)\, dx \tag{1.8-1}$$

By graphically plotting $f(x)$ versus x, the area under the curve is equal to the value of the integral.

1.8B Numerical Integration and Simpson's Rule

Often it is desirable or necessary to perform a numerical integration by computing the value of a definite integral from a set of numerical values of the integrand $f(x)$. This, of course, can be done graphically, but in most cases numerical methods suitable for the digital computer are desired.

The integral to be evaluated is Eq. (1.8-1), where the interval is $b - a$. The most generally used numerical method is the parabolic rule, often called *Simpson's rule*. This method divides the total interval $b - a$ into an even number of subintervals m, where

$$m = \frac{b - a}{h} \tag{1.8-2}$$

The value of h, a constant, is the spacing in x used. Then, approximating $f(x)$ by a parabola on each subinterval, Simpson's rule is

$$\int_{x=a}^{x=b} f(x)\, dx = \frac{h}{3}\,[f_o + 4(f_1 + f_3 + f_5 + \cdots + f_{m-1})$$

$$+ 2(f_2 + f_4 + f_6 + \cdots + f_{m-2}) + f_m] \tag{1.8-3}$$

where f_0 is the value of $f(x)$ at $x = a$; f_1 the value of $f(x)$ at $x = x_1, \ldots$; f_m the value of $f(x)$ at $x = b$. The reader should note that m must be an even number and the increments evenly spaced. This method is well suited for digital computation with a spreadsheet. Spreadsheets have numerical calculation methods built into their programs. Simpson's rule is a widely used numerical integration method.

In some cases the available experimental data for $f(x)$ are not at equally spaced increments of x. Then the numerical integration can be performed using the sum of the single-interval rectangles (trapezoidal rule) for the value of the interval. This is much less accurate than Simpson's rule. The trapezoidal-rule method becomes more accurate as the interval becomes smaller.

Sometimes the experimental data for $f(x)$ are spaced at large and/or irregular increments of x. These data can be smoothed by fitting a polynomial, exponential, logarithmic, or some other function to the data, which often can be integrated analytically. If the function is relatively complex, then it can be numerically integrated using Simpson's rule. All of this can be done using Excel and a spreadsheet.

PROBLEMS

1.2-1. *Temperature of a Chemical Process.* The temperature of a chemical reaction was found to be 353.2 K. What is the temperature in °F, °C, and °R?

Ans. 176°F, 80°C, 636°R

1.2-2. *Temperature for Smokehouse Processing of Meat.* In smokehouse processing of sausage meat, a final temperature of 155°F inside the sausage is often used. Calculate this temperature in °C, K, and °R.

1.3-1. *Molecular Weight of Air.* For purposes of most engineering calculations, air is assumed to be composed of 21 mol % oxygen and 79 mol % nitrogen. Calculate the average molecular weight.

Ans. 28.9 g mass/g mol, lb mass/lb mol, or kg mass/kg mol

1.3-2. *Oxidation of CO and Mole Units.* The gas CO is being oxidized by O_2 to form CO_2. How many kg of CO_2 will be formed from 56 kg of CO? Also, calculate the kg of O_2 theoretically needed for this reaction. (*Hint:* First write the balanced chemical equation to obtain the mol O_2 needed for 1.0 kg mol CO. Then calculate the kg mol of CO in 56 kg CO.)

Ans. 88.0 kg CO_2, 32.0 kg O_2

1.3-3. *Composition of a Gas Mixture.* A gaseous mixture contains 20 g of N_2, 83 g of O_2, and 45 g of CO_2. Calculate the composition in mole fraction and the average molecular weight of the mixture.

Ans. Average mol wt = 34.2 g mass/g mol, 34.2 kg mass/kg mol

1.3-4. *Composition of a Protein Solution.* A liquid solution contains 1.15 wt % of a protein, 0.27 wt % KCl, and the remainder water. The average molecular weight of the protein by gel permeation is 525 000 g mass/g mol. Calculate the mole fraction of each component in solution.

1.3-5. *Concentration of NaCl Solution.* An aqueous solution of NaCl has a concentration of 24.0 wt % NaCl with a density of 1.178 g/cm³ at 25°C. Calculate the following.
(a) Mole fraction of NaCl and water.
(b) Concentration of NaCl as g mol/liter, lb_m/ft³, lb_m/gal, and kg/m³.

1.4-1. *Conversion of Pressure Measurements in Freeze Drying.* In the experimental measurement of freeze drying of beef, an absolute pressure of 2.4 mm Hg was held in the chamber. Convert this pressure to atm, in. of water at 4°C, μm of Hg, and Pa. (*Hint:* See Appendix A.1 for conversion factors.)

Ans. 3.16×10^{-3} atm, 1.285 in. H_2O, 2400 μm Hg, 320 Pa

1.4-2. *Compression and Cooling of Nitrogen Gas.* A volume of 65.0 ft³ of N_2 gas at 90°F and 29.0 psig is compressed to 75 psig and cooled to 65°F. Calculate the final volume in ft³ and the final density in lb_m/ft³. [*Hint:* Be sure to convert all pressures to psia first and then to atm. Substitute original conditions into Eq. (1.4-1) to obtain *n*, lb mol.]

1.4-3. *Gas Composition and Volume.* A gas mixture of 0.13 g mol NH_3, 1.27 g mol N_2, and 0.025 g mol H_2O vapor is contained at a total pressure of 830 mm Hg and 323 K. Calculate the following.
(a) Mole fraction of each component.
(b) Partial pressure of each component in mm Hg.
(c) Total volume of mixture in m³ and ft³.

1.4-4. *Evaporation of a Heat-Sensitive Organic Liquid.* An organic liquid is being evaporated from a liquid solution containing a few percent nonvolatile dissolved solids. Since it is heat-sensitive and may discolor at high temperatures, it will be evaporated under vacuum. If the lowest absolute pressure that can be obtained in the apparatus is 12.0 mm Hg, what will be the temperature of evaporation in K? It will be assumed that the small amount of solids does not affect the vapor pressure, which is given as follows:

$$\log P_A = -2250\left(\frac{1}{T}\right) + 9.05$$

where P_A is in mm Hg and T in K.

Ans. $T = 282.3$ K or 9.1°C

1.5-1. *Evaporation of Cane Sugar Solutions.* An evaporator is used to concentrate cane sugar solutions. A feed of 10 000 kg/d of a solution containing 38 wt % sugar is evaporated, producing a 74 wt % solution. Calculate the weight of solution produced and amount of water removed.

Ans.   5135 kg/d of 74 wt % solution, 4865 kg/d water

1.5-2. *Processing of Fish Meal.* Fish are processed into fish meal and used as a supplementary protein food. In the processing the oil is first extracted to produce wet fish cake containing 80 wt % water and 20 wt % bone-dry cake. This wet cake feed is dried in rotary-drum dryers to give a "dry" fish cake product containing 40 wt % water. Finally, the product is finely ground and packed. Calculate the kg/h of wet cake feed needed to produce 1000 kg/h of "dry" fish cake product.

Ans.   3000 kg/h wet cake feed

1.5-3. *Drying of Lumber.* A batch of 100 kg of wet lumber containing 11 wt % moisture is dried to a water content of 6.38 kg water/1.0 kg bone-dry lumber. What is the weight of "dried" lumber and the amount of water removed?

1.5-4. *Processing of Paper Pulp.* A wet paper pulp contains 68 wt % water. After the pulp was dried, it was found that 55% of the original water in the wet pulp was removed. Calculate the composition of the "dried" pulp and its weight for a feed of 1000 kg/min of wet pulp.

1.5-5. *Production of Jam from Crushed Fruit in Two Stages.* In a process producing jam (C1), crushed fruit containing 14 wt % soluble solids is mixed in a mixer with sugar (1.22 kg sugar/1.00 kg crushed fruit) and pectin (0.0025 kg pectin/1.00 kg crushed fruit). The resultant mixture is then evaporated in a kettle to produce a jam containing 67 wt % soluble solids. For a feed of 1000 kg crushed fruit, calculate the kg mixture from the mixer, kg water evaporated, and kg jam produced.

Ans.   2222.5 kg mixture, 189 kg water, 2033.5 kg jam

1.5-6. *Drying of Cassava (Tapioca) Root.* Tapioca flour is used in many countries for bread and similar products. The flour is made by drying coarse granules of the cassava root containing 66 wt % moisture to 5% moisture and then grinding to produce a flour. How many kg of granules must be dried and how much water removed to produce 5000 kg/h of flour?

1.5-7. *Processing of Soybeans in Three Stages.* A feed of 10 000 kg of soybeans is processed in a sequence of three stages or steps (E1). The feed contains 35 wt % protein, 27.1 wt % carbohydrate, 9.4 wt % fiber and ash, 10.5 wt % moisture, and 18.0 wt % oil. In the first stage the beans are crushed and pressed to remove oil, giving an expressed-oil stream and a stream of pressed beans containing 6% oil. Assume no loss of other constituents with the oil stream. In the second step the pressed beans are extracted with hexane to produce an extracted-meal stream containing 0.5 wt % oil and a hexane–oil stream. Assume no hexane in the extracted meal. Finally, in the last step the extracted meal is dried to give a dried meal of 8 wt % moisture. Calculate:

(a) Kg of pressed beans from the first stage.

(b) Kg of extracted meal from stage 2.

(c) Kg of final dried meal and the wt % protein in the dried meal.

Ans. (a) 8723 kg; (b) 8241 kg; (c) 7816 kg, 44.8 wt % protein

1.5-8. *Recycle in a Dryer.* A solid material containing 15.0 wt % moisture is dried so that it contains 7.0 wt % water by blowing fresh warm air mixed with recycled air over the solid in the dryer. The inlet fresh air has a humidity of 0.01 kg water/kg dry air, the air from the drier that is recycled has a humidity of 0.1 kg water/kg dry air, and the mixed air to the dryer, 0.03 kg water/kg dry air. For a feed of 100 kg solid/h fed to the dryer, calculate the kg dry air/h in the fresh air, the kg dry air/h in the recycled air, and the kg/h of "dried" product.

Ans. 95.6 kg/h dry air in fresh air, 27.3 kg/h dry air in recycled air, and 91.4 kg/h "dried" product

1.5-9. *Crystallization and Recycle.* It is desired to produce 1000 kg/h of $Na_3PO_4 \cdot 12H_2O$ crystals from a feed solution containing 5.6 wt % Na_3PO_4 and traces of impurity. The original solution is first evaporated in an evaporator to a 35 wt % Na_3PO_4 solution and then cooled to 293 K in a crystallizer, where the hydrated crystals and a mother-liquor solution are removed. One out of every 10 kg of mother liquor is discarded to waste to get rid of the impurities, and the remaining mother liquor is recycled to the evaporator. The solubility of Na_3PO_4 at 293 K is 9.91 wt %. Calculate the kg/h of feed solution and kg/h of water evaporated.

Ans. 7771 kg/h feed, 6739 kg/h water

1.5-10. *Evaporation and Bypass in Orange Juice Concentration.* In a process for concentrating 1000 kg of freshly extracted orange juice (C1) containing 12.5 wt % solids, the juice is strained, yielding 800 kg of strained juice and 200 kg of pulpy juice. The strained juice is concentrated in a vacuum evaporator to give an evaporated juice of 58% solids. The 200 kg of pulpy juice is bypassed around the evaporator and mixed with the evaporated juice in a mixer to improve the flavor. This final concentrated juice contains 42 wt % solids. Calculate the concentration of solids in the strained juice, the kg of final concentrated juice, and the concentration of solids in the pulpy juice bypassed. (*Hint:* First, make a total balance and then a solids balance on the overall process. Next, make a balance on the evaporator. Finally, make a balance on the mixer.)

Ans. 34.2 wt % solids in pulpy juice

1.5-11. *Manufacture of Acetylene.* For the making of 6000 ft³ of acetylene (CHCH) gas at 70°F and 750 mm Hg, solid calcium carbide (CaC_2) which contains 97 wt % CaC_2 and 3 wt % solid inerts is used along with water. The reaction is

$$CaC_2 + 2H_2O \rightarrow CHCH + Ca(OH)_2\downarrow$$

The final lime slurry contains water, solid inerts, and $Ca(OH)_2$ lime. In this slurry the total wt % solids of inerts plus $Ca(OH)_2$ is 20%. How many lb of water must be added and how many lb of final lime slurry is produced? [*Hint:* Use a basis of 6000 ft³ and convert to lb mol. This gives 15.30 lb mol C_2H_2, 15.30 lb mol $Ca(OH)_2$, and 15.30 lb mol CaC_2 added. Convert lb mol CaC_2 feed to lb and calculate lb inerts added. The total lb solids in the slurry is then the sum of the $Ca(OH)_2$ plus inerts. In calculating the water added, remember that some is consumed in the reaction.]

Ans. 5200 lb water added (2359 kg), 5815 lb lime slurry (2638 kg)

1.5-12. *Combustion of Solid Fuel.* A fuel analyzes 74.0 wt % C and 12.0% ash (inert). Air is added to burn the fuel, producing a flue gas of 12.4% CO_2, 1.2% CO, 5.7% O_2, and

80.7% N_2. Calculate the kg of fuel used for 100 kg mol of outlet flue gas and the kg mol of air used. (*Hint:* First calculate the mol O_2 added in the air, using the fact that the N_2 in the flue gas equals the N_2 added in the air. Then make a carbon balance to obtain the total moles of C added.)

1.5-13. *Burning of Coke.* A furnace burns a coke containing 81.0 wt % C, 0.8% H, and the rest inert ash. The furnace uses 60% excess air (air over and above that needed to burn all C to CO_2 and H to H_2O). Calculate the moles of all components in the flue gas if only 95% of the carbon goes to CO_2 and 5% to CO.

1.5-14. *Production of Formaldehyde.* Formaldehyde (CH_2O) is made by the catalytic oxidation of pure methanol vapor and air in a reactor. The moles from this reactor are 63.1 N_2, 13.4 O_2, 5.9 H_2O, 4.1 CH_2O, 12.3 CH_3OH, and 1.2 HCOOH. The reaction is

$$CH_3OH + \tfrac{1}{2}O_2 \rightarrow CH_2O + H_2O$$

A side reaction occurring is

$$CH_2O + \tfrac{1}{2}O_2 \rightarrow HCOOH$$

Calculate the mol methanol feed, mol air feed, and percent conversion of methanol to formaldehyde.

Ans. 17.6 mol CH_3OH, 79.8 mol air, 23.3% conversion

1.6-1. *Heating of CO_2 Gas.* A total of 250 g of CO_2 gas at 373 K is heated to 623 K at 101.32 kPa total pressure. Calculate the amount of heat needed in cal, btu, and kJ.

Ans. 15 050 cal, 59.7 btu, 62.98 kJ

1.6-2. *Heating a Gas Mixture.* A mixture of 25 lb mol N_2 and 75 lb mol CH_4 is being heated from 400°F to 800°F at 1 atm pressure. Calculate the total amount of heat needed in btu.

1.6-3. *Final Temperature in Heating Applesauce.* A mixture of 454 kg of applesauce at 10°C is heated in a heat exchanger by adding 121 300 kJ. Calculate the outlet temperature of the applesauce. (*Hint:* In Appendix A.4 a heat capacity for applesauce is given at 32.8°C. Assume that this is constant and use this as the average c_{pm}.)

Ans. 76.5°C

1.6-4. *Use of Steam Tables.* Using the steam tables, determine the enthalpy change for 1 lb water for each of the following cases:
(a) Heating liquid water from 40°F to 240°F at 30 psia. (Note that the effect of total pressure on the enthalpy of liquid water can be neglected.)
(b) Heating liquid water from 40°F to 240°F and vaporizing at 240°F and 24.97 psia.
(c) Cooling and condensing a saturated vapor at 212°F and 1 atm abs to a liquid at 60°F.
(d) Condensing a saturated vapor at 212°F and 1 atm abs.

Ans. (a) 200.42 btu/lb$_m$; (b) 1152.7 btu/lb$_m$; (c) −1122.4 btu/lb$_m$; (d) −970.3 btu/lb$_m$, −2256.9 kJ/kg

1.6-5. *Heating and Vaporization Using Steam Tables.* A flow rate of 1000 kg/h of water at 21.1°C is heated to 110°C when the total pressure is 244.2 kPa in the first stage of a process. In the second stage at the same pressure the water is heated further, until it is all vaporized at its boiling point. Calculate the total enthalpy change in the first stage and in both stages.

1.6-6. *Combustion of CH₄ and H₂.* For 100 g mol of a gas mixture of 75 mol % CH_4 and 25% H_2, calculate the total heat of combustion of the mixture at 298 K and 101.32 kPa, assuming that combustion is complete.

1.6-7. *Heat of Reaction from Heats of Formation.* For the reaction

$$4NH_3(g) + 5O_2(g) \rightarrow 4NO(g) + 6H_2O(g)$$

calculate the heat of reaction, ΔH, at 298 K and 101.32 kPa for 4 g mol of NH_3 reacting.

Ans. ΔH, heat of reaction $= -904.7$ kJ

1.7-1. *Heat Balance and Cooling of Milk.* In the processing of rich cows' milk, 4540 kg/h of milk is cooled from 60°C to 4.44°C by a refrigerant. Calculate the heat removed from the milk.

Ans. Heat removed $= 269.6$ kW

1.7-2. *Heating of Oil by Air.* A flow of 2200 lb_m/h of hydrocarbon oil at 100°F enters a heat exchanger, where it is heated to 150°F by hot air. The hot air enters at 300°F and is to leave at 200°F. Calculate the total lb mol air/h needed. The mean heat capacity of the oil is 0.45 btu/$lb_m \cdot$ °F.

Ans. 70.1 lb mol air/h, 31.8 kg mol/h

1.7-3. *Combustion of Methane in a Furnace.* A gas stream of 10 000 kg mol/h of CH_4 at 101.32 kPa and 373 K is burned in a furnace using air at 313 K. The combustion is complete and 50% excess air is used. The flue gas leaves the furnace at 673 K. Calculate the heat removed in the furnace. (*Hint:* Use a datum of 298 K and liquid water at 298 K. The input items will be the following: the enthalpy of CH_4 at 373 K referred to 298 K; the enthalpy of the air at 313 K referred to 298 K; $-\Delta H_c^0$, the heat of combustion of CH_4 at 298 K, which is referred to liquid water; and q, the heat added. The output items will include: the enthalpies of CO_2, O_2, N_2, and H_2O gases at 673 K referred to 298 K; and the latent heat of H_2O vapor at 298 K and 101.32 kPa from Appendix A.2. It is necessary to include this latent heat since the basis of the calculation and of the ΔH_c^0 is liquid water.)

1.7-4. *Preheating Air by Steam for Use in a Dryer.* An air stream at 32.2°C is to be used in a dryer and is first preheated in a steam heater, where it is heated to 65.5°C. The air flow is 1000 kg mol/h. The steam enters the heater saturated at 148.9°C, is condensed and cooled, and leaves as a liquid at 137.8°C. Calculate the amount of steam used in kg/h.

Ans. 450 kg steam/h

1.7-5. *Cooling of Cans of Potato Soup After Thermal Processing.* A total of 1500 cans of potato soup undergo thermal processing in a retort at 240°F. The cans are then cooled to 100°F in the retort before being removed from the retort by cooling water, which enters at 75°F and leaves at 85°F. Calculate the lb of cooling water needed. Each can of soup contains 1.0 lb of liquid soup, and the empty metal can weighs 0.16 lb. The mean heat capacity of the soup is 0.94 btu/$lb_m \cdot$ °F and that of the metal can is 0.12 btu/$lb_m \cdot$ °F. A metal rack or basket which is used to hold the cans in the retort weighs 350 lb and has a heat capacity of 0.12 btu/$lb_m \cdot$ °F. Assume that the metal rack is cooled from 240°F to 85°F, the temperature of the outlet water. The amount of heat removed from the retort walls in cooling from 240 to 100°F is 10 000 btu. Radiation loss from the retort during cooling is estimated as 5000 btu.

Ans. 21 320 lb water, 9670 kg

1.8-1. *Numerical Integration Using Simpson's Method.* The following experimental data for $y = f(x)$ were obtained:

x	$f(x)$	x	$f(x)$
0	100	0.4	53
0.1	75	0.5	60
0.2	60.5	0.6	72.5
0.3	53.5		

Determine the integral using Simpson's method:

$$A = \int_{x=0}^{x=0.6} f(x)\, dx$$

Ans. $A = 38.45$

1.8-2. *Numerical Integration to Obtain Wastewater Flow.* The rate of flow of wastewater in an open channel has been measured and the following data obtained:

Time (min)	Flow (m³/min)	Time (min)	Flow (m³/min)
0	655	70	800
10	705	80	725
20	780	90	670
30	830	100	640
40	870	110	620
50	890	120	610
60	870		

Determine the total flow in m³ for 120 min by numerical integration.

Ans. 92 350 m³

REFERENCES

(C1) CHARM, S. E. *The Fundamentals of Food Engineering,* 2nd ed. Westport, Conn.: Avi Publishing Co., Inc., 1971.

(E1) EARLE, R. L. *Unit Operations in Food Processing.* Oxford: Pergamon Press, Inc., 1966.

(H1) HOUGEN, O. A., WATSON, K. M., and RAGATZ, R. A. *Chemical Process Principles,* Part I, 2nd ed. New York: John Wiley & Sons, Inc., 1954.

(O1) OKOS, M. R. M.S. thesis. Ohio State University, Columbus, Ohio, 1972.

(P1) PERRY, R. H., and GREEN, D. *Perry's Chemical Engineers' Handbook,* 6th ed. New York: McGraw-Hill Book Company, 1984.

(S1) SOBER, H. A. *Handbook of Biochemistry, Selected Data for Molecular Biology,* 2nd ed. Boca Raton, Fla.: Chemical Rubber Co., Inc., 1970.

(W1) WEAST, R. C., and SELBY, S. M. *Handbook of Chemistry and Physics,* 48th ed. Boca Raton, Fla.: Chemical Rubber Co., Inc., 1967–1968.

CHAPTER 2

Principles of Momentum Transfer and Overall Balances

2.1 INTRODUCTION

The flow and behavior of fluids is important in many of the separation processes in process engineering. A fluid may be defined as a substance that does not permanently resist distortion and, hence, will change its shape. In this text gases, liquids, and vapors are considered to have the characteristics of fluids and to obey many of the same laws.

In the process industries, many of the materials are in fluid form and must be stored, handled, pumped, and processed, so it is necessary that we become familiar with the principles that govern the flow of fluids as well as with the equipment used. Typical fluids encountered include water, air, CO_2, oil, slurries, and thick syrups.

If a fluid is inappreciably affected by changes in pressure, it is said to be *incompressible.* Most liquids are incompressible. Gases are considered to be *compressible* fluids. However, if gases are subjected to small percentage changes in pressure and temperature, their density changes will be small and they can be considered to be incompressible.

Like all physical matter, a fluid is composed of an extremely large number of molecules per unit volume. A theory such as the kinetic theory of gases or statistical mechanics treats the motions of molecules in terms of statistical groups and not in terms of individual molecules. In engineering we are mainly concerned with the bulk or macroscopic behavior of a fluid rather than the individual molecular or microscopic behavior.

In momentum transfer we treat the fluid as a continuous distribution of matter, or a "continuum." This treatment as a continuum is valid when the smallest volume of fluid contains a number of molecules large enough that a statistical average is meaningful and the macroscopic properties of the fluid, such as density, pressure, and so on, vary smoothly or continuously from point to point.

The study of *momentum transfer,* or *fluid mechanics* as it is often called, can be divided into two branches: *fluid statics,* or fluids at rest, and *fluid dynamics,* or fluids in motion. In Section 2.2 we treat fluid statics; in the remaining sections of Chapter 2 and in Chapter 3, fluid dynamics. Since in fluid dynamics momentum is being transferred, the term "momentum transfer" or "transport" is usually used. In Section 2.3 momentum transfer is related to heat and mass transfer.

2.2   FLUID STATICS

2.2A   Force, Units, and Dimensions

In a static fluid an important property is the pressure in the fluid. Pressure is familiar as a surface force exerted by a fluid against the walls of its container. Also, pressure exists at any point in a volume of a fluid.

In order to understand *pressure*, which is defined as force exerted per unit area, we must first discuss a basic law of Newton's. This equation for calculation of the force exerted by a mass under the influence of gravity is

$$F = mg \qquad \text{(SI units)}$$

$$F = \frac{mg}{g_c} \qquad \text{(English units)}$$

(2.2-1)

where in SI units F is the force exerted in newtons N $(\text{kg} \cdot \text{m/s}^2)$, m the mass in kg, and g the standard acceleration of gravity, 9.80665 m/s^2.

In English units, F is in lb$_f$, m in lb$_m$, g is 32.1740 ft/s^2, and g_c (a gravitational conversion factor) is 32.174 lb$_m \cdot$ ft/lb$_f \cdot$ s^2. The use of the conversion factor g_c means that g/g_c has a value of 1.0 lb$_f$/lb$_m$ and that 1 lb$_m$ conveniently gives a force equal to 1 lb$_f$. Often when units of pressure are given, the word "force" is omitted, as in lb/in.^{2} (psi) instead of lb$_f$/in.^{2}. When the mass m is given in g mass, F is g force, $g = 980.665$ cm/s^2, and $g_c = 980.665$ g mass $\cdot$ cm/g force $\cdot$ s^2. However, the units g force are seldom used.

Another system of units sometimes used in Eq. (2.2-1) is that where the g_c is omitted and the force $(F = mg)$ is given as lb$_m \cdot$ ft/s^2, called *poundals*. Then 1 lb$_m$ acted on by gravity will give a force of 32.174 poundals (lb$_m \cdot$ ft/s^2). Or if 1 g mass is used, the force $(F = mg)$ is expressed in terms of dynes (g $\cdot$ cm/s^2). This is the centimeter–gram–second (cgs) systems of units.

Conversion factors for different units of force and of force per unit area (pressure) are given in Appendix A.1. Note that always in the SI system, and usually in the cgs system, the term g_c is not used.

EXAMPLE 2.2-1.   Units and Dimensions of Force
Calculate the force exerted by 3 lb mass in terms of the following:
      (a) Lb force (English units).
      (b) Dynes (cgs units).
      (c) Newtons (SI units).

Solution: For part (a), using Eq. (2.2-1),

$$F \text{ (force)} = m\frac{g}{g_c} = (3 \text{ lb}_m)\left(32.174 \frac{\text{ft}}{\text{s}^2}\right)\left(\frac{1}{32.174 \dfrac{\text{lb}_m \cdot \text{ft}}{\text{lb}_f \cdot \text{s}^2}}\right) = 3 \text{ lb force (lb}_f)$$

For part (b),

$$F = mg = (3 \text{ lb}_m)\left(453.59 \frac{\text{g}}{\text{lb}_m}\right)\left(980.665 \frac{\text{cm}}{\text{s}^2}\right)$$

$$= 1.332 \times 10^6 \frac{\text{g} \cdot \text{cm}}{\text{s}^2} = 1.332 \times 10^6 \text{ dyn}$$

As an alternative method for part (b), from Appendix A.1,

$$1 \text{ dyn} = 2.2481 \times 10^{-6} \text{ lb}_f$$

$$F = (3 \text{ lb}_f) \left(\frac{1}{2.2481 \times 10^{-6} \text{ lb}_f/\text{dyn}} \right) = 1.332 \times 10^6 \text{ dyn}$$

To calculate newtons in part (c),

$$F = mg = \left(3 \text{ lb}_m \times \frac{1 \text{ kg}}{2.2046 \text{ lb}_m} \right) \left(9.80665 \frac{\text{m}}{\text{s}^2} \right)$$

$$= 13.32 \frac{\text{kg} \cdot \text{m}}{\text{s}^2} = 13.32 \text{ N}$$

As an alternative method, using values from Appendix A.1,

$$1 \frac{\text{g} \cdot \text{cm}}{\text{s}^2} \text{ (dyn)} = 10^{-5} \frac{\text{kg} \cdot \text{m}}{\text{s}^2} \text{ (newton)}$$

$$F = (1.332 \times 10^6 \text{ dyn}) \left(10^{-5} \frac{\text{newton}}{\text{dyn}} \right) = 13.32 \text{ N}$$

2.2B Pressure in a Fluid

Since Eq. (2.2-1) gives the force exerted by a mass under the influence of gravity, the force exerted by a mass of fluid on a supporting area, or force/unit area (pressure), also follows from this equation. In Fig. 2.2-1 a stationary column of fluid of height h_2 m and constant cross-sectional area A m^2, where $A = A_0 = A_1 = A_2$, is shown. The pressure above the fluid is P_0 N/m^2; that is, this could be the pressure of the atmosphere above the fluid. The fluid at any point, say h_1, must support all the fluid above it. It can be shown that the forces at any given point in a nonmoving or static fluid must be the same in all directions. Also, for a fluid at rest, the force/unit area, or pressure, is the same at all points with the same elevation. For example, at h_1 m from the top, the pressure is the same at all points shown on the cross-sectional area A_1.

FIGURE 2.2-1.   *Pressure in a static fluid.*

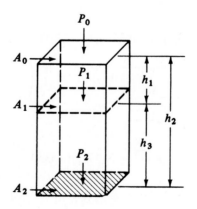

Chapter 2   Principles of Momentum Transfer and Overall Balances

The use of Eq. (2.2-1) will be shown in calculating the pressure at different vertical points in Fig. 2.2-1. The total mass of fluid for h_2 m height and density ρ kg/m^3 is

$$\text{total kg fluid} = (h_2\text{ m})(A\text{ m}^2)\left(\rho\,\frac{\text{kg}}{\text{m}^3}\right) = h_2 A\rho\text{ kg} \tag{2.2-2}$$

Substituting into Eq. (2.2-2), the total force F of the fluid on area A_1 due to the fluid only is

$$F = (h_2 A\rho\text{ kg})(g\text{ m/s}^2) = h_2 A\rho g\,\frac{\text{kg}\cdot\text{m}}{\text{s}^2}\,(\text{N}) \tag{2.2-3}$$

The pressure P is defined as force/unit area:

$$P = \frac{F}{A} = (h_2 A\rho g)\frac{1}{A} = h_2\rho g\text{ N/m}^2 \quad \text{or} \quad \text{Pa} \tag{2.2-4}$$

This is the pressure on A_2 due to the mass of the fluid above it. However, to get the total pressure P_2 on A_2, the pressure P_0 on the top of the fluid must be added:

$$P_2 = h_2\rho g + P_0\text{ N/m}^2 \quad \text{or} \quad \text{Pa} \tag{2.2-5}$$

Equation (2.2-5) is the fundamental equation for calculating the pressure in a fluid at any depth. To calculate P_1,

$$P_1 = h_1\rho g + P_0 \tag{2.2-6}$$

The pressure difference between points 2 and 1 is

$$P_2 - P_1 = (h_2\rho g + P_0) - (h_1\rho g + P_0) = (h_2 - h_1)\rho g \quad \text{(SI units)}$$

$$P_2 - P_1 = (h_2 - h_1)\rho\,\frac{g}{g_c} \quad \text{(English units)} \tag{2.2-7}$$

Since it is the vertical height of a fluid that determines the pressure in a fluid, the shape of the vessel does not affect the pressure. For example, in Fig. 2.2-2, the pressure P_1 at the bottom of all three vessels is the same and is equal to $h_1\rho g + P_0$.

EXAMPLE 2.2-2.   *Pressure in Storage Tank*
A large storage tank contains oil having a density of 917 kg/m^3 (0.917 g/cm^3). The tank is 3.66 m (12.0 ft) tall and is vented (open) to the atmosphere of 1 atm abs at the top. The tank is filled with oil to a depth of 3.05 m (10 ft) and also con-

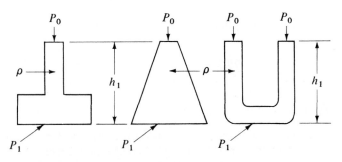

FIGURE 2.2-2.   *Pressure in vessels of various shapes.*

FIGURE 2.2-3. *Storage tank in Example 2.2-2.*

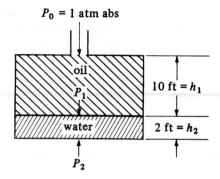

tains 0.61 m (2.0 ft) of water in the bottom of the tank. Calculate the pressure in Pa and psia 3.05 m from the top of the tank and at the bottom. Also calculate the gage pressure at the tank bottom.

Solution: First a sketch is made of the tank, as shown in Fig. 2.2-3. The pressure $P_0 = 1$ atm abs $= 14.696$ psia (from Appendix A.1). Also,

$$P_0 = 1.01325 \times 10^5 \text{ Pa}$$

From Eq. (2.2-6), using English and then SI units,

$$P_1 = h_1 \rho_{\text{oil}} \frac{g}{g_c} + P_0 = (10 \text{ ft}) \left(0.917 \times 62.43 \frac{\text{lb}_\text{m}}{\text{ft}^3} \right) \left(1.0 \frac{\text{lb}_\text{f}}{\text{lb}_\text{m}} \right) \left(\frac{1}{144 \text{ in.}^2/\text{ft}^2} \right)$$
$$+ 14.696 \text{ lb}_\text{f}/\text{in.}^2 = 18.68 \text{ psia}$$

$$P_1 = h_1 \rho_{\text{oil}} g + P_0 = (3.05 \text{ m}) \left(917 \frac{\text{kg}}{\text{m}^3} \right) \left(9.8066 \frac{\text{m}}{\text{s}^2} \right) + 1.0132 \times 10^5$$

$$= 1.287 \times 10^5 \text{ Pa}$$

To calculate P_2 at the bottom of the tank, $\rho_{\text{water}} = 1.00 \text{ g/cm}^3$ and

$$P_2 = h_2 \rho_{\text{water}} \frac{g}{g_c} + P_1 = (2.0)(1.00 \times 62.43)(1.0)(\tfrac{1}{144}) + 18.68$$

$$= 19.55 \text{ psia}$$

$$= h_2 \rho_{\text{water}} g + P_1 = (0.61)(1000)(9.8066) + 1.287 \times 10^5$$

$$= 1.347 \times 10^5 \text{ Pa}$$

The gage pressure at the bottom is equal to the absolute pressure P_2 minus 1 atm pressure:

$$P_{\text{gage}} = 19.55 \text{ psia} - 14.696 \text{ psia} = 4.85 \text{ psig}$$

2.2C Head of a Fluid

Pressures are given in many different sets of units, such as psia, dyn/cm², and newtons/m², as given in Appendix A.1. However, a common method of expressing pressures is in terms of head in m or feet of a particular fluid. This height or head in m or feet of the given fluid will exert the same pressure as the pressures it represents. Using Eq. (2.2-4), which relates pressure P and height h of a fluid, and solving for h, which is the head in m,

$$h(\text{head}) = \frac{P}{\rho g} \text{ m} \quad (\text{SI})$$

$$h = \frac{P g_c}{\rho g} \text{ ft} \quad (\text{English}) \tag{2.2-8}$$

EXAMPLE 2.2-3. *Conversion of Pressure to Head of a Fluid*

Given the pressure of 1 standard atm as 101.325 kN/m^2 (Appendix A.1), do as follows:

(a) Convert this pressure to head in m water at 4°C.
(b) Convert this pressure to head in m Hg at 0°C.

Solution: For part (a), the density of water at 4°C in Appendix A.2 is 1.000 g/cm^3. From A.1, a density of 1.000 g/cm^3 equals 1000 kg/m^3. Substituting these values into Eq. (2.2-8),

$$h(\text{head}) = \frac{P}{\rho g} = \frac{101.325 \times 10^3}{(1000)(9.80665)}$$

$$= 10.33 \text{ m of water at 4°C}$$

For part (b), the density of Hg in Appendix A.1 is 13.5955 g/cm^3. For equal pressures P from different fluids, Eq. (2.2-8) can be rewritten as

$$P = \rho_{\text{Hg}} h_{\text{Hg}} g = \rho_{\text{H}_2\text{O}} h_{\text{H}_2\text{O}} g \tag{2.2-9}$$

Solving for h_{Hg} in Eq. (2.2-9) and substituting known values,

$$h_{\text{Hg}}(\text{head}) = \frac{\rho_{\text{H}_2\text{O}}}{\rho_{\text{Hg}}} h_{\text{H}_2\text{O}} = \left(\frac{1.000}{13.5955} \right)(10.33) = 0.760 \text{ m Hg}$$

2.2D Devices to Measure Pressure and Pressure Differences

In chemical and other industrial processing plants, it is often important to measure and control the pressure in a vessel or process and/or the liquid level in a vessel. Also, since many fluids are flowing in a pipe or conduit, it is necessary to measure the rate at which the fluid is flowing. Many of these flow meters depend upon devices for measuring a pressure or pressure difference. Some common devices are considered in the following paragraphs.

1. Simple U-tube manometer. The U-tube manometer is shown in Fig. 2.2-4a. The pressure p_a N/m^2 is exerted on one arm of the U tube and p_b on the other arm. Both pressures p_a and p_b could be pressure taps from a fluid meter, or p_a could be a pressure tap and p_b the atmospheric pressure. The top of the manometer is filled with liquid B, having a density of ρ_B kg/m^3, and the bottom with a more dense fluid A, having a density of ρ_A kg/m^3. Liquid A is immiscible with B. To derive the relationship between p_a and p_b, p_a is the pressure at point 1 and p_b at point 5. The pressure at point 2 is

$$p_2 = p_a + (Z + R)\rho_B g \text{ N/m}^2 \tag{2.2-10}$$

where R is the reading of the manometer in m. The pressure at point 3 must be equal to that at 2 by the principles of hydrostatics:

$$p_3 = p_2 \qquad \textbf{(2.2-11)}$$

The pressure at point 3 also equals the following:

$$p_3 = p_b + Z\rho_B g + R\rho_A g \qquad \textbf{(2.2-12)}$$

Equating Eq. (2.2-10) to (2.2-12) and solving,

$$p_a + (Z + R)\rho_B g = p_b + Z\rho_B g + R\rho_A g \qquad \textbf{(2.2-13)}$$

$$p_a - p_b = R(\rho_A - \rho_B)g \qquad \text{(SI)}$$

$$p_a - p_b = R(\rho_A - \rho_B)\frac{g}{g_c} \qquad \text{(English)} \qquad \textbf{(2.2-14)}$$

The reader should note that the distance Z does not enter into the final result nor do the tube dimensions, provided that p_a and p_b are measured in the same horizontal plane.

EXAMPLE 2.2-4. Pressure Difference in a Manometer

A manometer, as shown in Fig. 2.2-4a, is being used to measure the head or pressure drop across a flow meter. The heavier fluid is mercury, with a density of 13.6 g/cm³, and the top fluid is water, with a density of 1.00 g/cm³. The reading on the manometer is $R = 32.7$ cm. Calculate the pressure difference in N/m² using SI units.

Solution: Converting R to m,

$$R = \frac{32.7}{100} = 0.327 \text{ m}$$

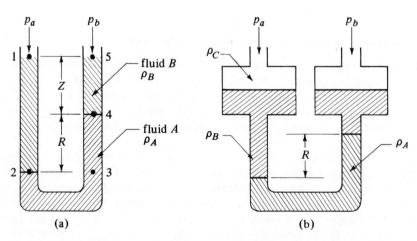

(a) (b)

FIGURE 2.2-4. *Manometers to measure pressure differences: (a) U tube; (b) two-fluid U tube.*

Also converting ρ_A and ρ_B to kg/m^3 and substituting into Eq. (2.2-14),

$$p_a - p_b = R(\rho_A - \rho_B)g = (0.327 \text{ m})[(13.6 - 1.0)(1000 \text{ kg/m}^3)](9.8066 \text{ m/s}^2)$$

$$= 4.040 \times 10^4 \text{ N/m}^2 \text{ (5.85 psia)}$$

2. Two-fluid U tube. In Fig. 2.2-4b a two-fluid U tube is shown, which is a sensitive device for measuring small heads or pressure differences. Let A m^2 be the cross-sectional area of each of the large reservoirs and a m^2 be the cross-sectional area of each of the tubes forming the U. Proceeding and making a pressure balance as for the U tube,

$$p_a - p_b = (R - R_0)\left(\rho_A - \rho_B + \frac{a}{A}\rho_B - \frac{a}{A}\rho_C\right)g \qquad \textbf{(2.2-15)}$$

where R_0 is the reading when $p_a = p_b$, R is the actual reading, ρ_A is the density of the heavier fluid, and ρ_B is the density of the lighter fluid. Usually, a/A is made sufficiently small as to be negligible, and also R_0 is often adjusted to zero; then

$$p_a - p_b = R(\rho_A - \rho_B)g \qquad \text{(SI)}$$

$$p_a - p_b = R(\rho_A - \rho_B)\frac{g}{g_c} \qquad \text{(English)} \qquad \textbf{(2.2-16)}$$

If ρ_A and ρ_B are close to each other, the reading R is magnified.

EXAMPLE 2.2-5. Pressure Measurement in a Vessel

The U-tube manometer in Fig. 2.2-5a is used to measure the pressure p_A in a vessel containing a liquid with a density ρ_A. Derive the equation relating the pressure p_A and the reading on the manometer as shown.

Solution: At point 2 the pressure is

$$p_2 = p_{\text{atm}} + h_2\rho_B g \text{ N/m}^2 \qquad \textbf{(2.2-17)}$$

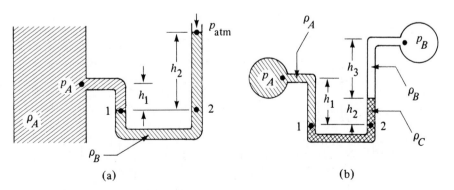

(a) (b)

FIGURE 2.2-5. *Measurements of pressure in vessels: (a) measurement of pressure in a vessel, (b) measurement of differential pressure.*

At point 1 the pressure is

$$p_1 = p_A + h_1 \rho_A g \tag{2.2-18}$$

Equating $p_1 = p_2$ by the principles of hydrostatics and rearranging,

$$p_A = p_{\text{atm}} + h_2 \rho_B g - h_1 \rho_A g \tag{2.2-19}$$

Another example of a U-tube manometer is shown in Fig. 2.2-5b. In this case the device is used to measure the pressure difference between two vessels.

3. Bourdon pressure gage. Although manometers are used to measure pressures, the most common pressure-measuring device is the mechanical Bourdon-tube pressure gage. A coiled hollow tube in the gage tends to straighten out when subjected to internal pressure, and the degree of straightening depends on the pressure difference between the inside and outside pressures. The tube is connected to a pointer on a calibrated dial.

4. Gravity separator for two immiscible liquids. In Fig. 2.2-6 a continuous gravity separator (decanter) is shown for the separation of two immiscible liquids A (heavy liquid) and B (light liquid). The feed mixture of the two liquids enters at one end of the separator vessel, and the liquids flow slowly to the other end and separate into two distinct layers. Each liquid flows through a separate overflow line as shown. Assuming the frictional resistance to the flow of the liquids is essentially negligible, the principles of fluid statics can be used to analyze the performance.

In Fig. 2.2-6, the depth of the layer of heavy liquid A is h_{A1} m and that of B is h_B. The total depth $h_T = h_{A1} + h_B$ and is fixed by position of the overflow line for B. The heavy liquid A discharges through an overflow leg h_{A2} m above the vessel bottom. The vessel and the overflow lines are vented to the atmosphere. A hydrostatic balance gives

$$h_B \rho_B g + h_{A1} \rho_A g = h_{A2} \rho_A g \tag{2.2-20}$$

Substituting $h_B = h_T - h_{A1}$ into Eq. (2.2-20) and solving for h_{A1},

$$h_{A1} = \frac{h_{A2} - h_T \rho_B / \rho_A}{1 - \rho_B / \rho_A} \tag{2.2-21}$$

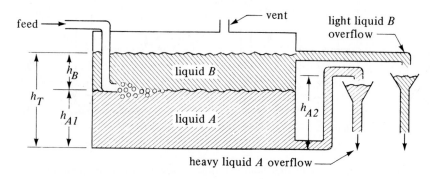

FIGURE 2.2-6. *Continuous atmospheric gravity separator for immiscible liquids.*

This shows that the position of the interface or height h_{A1} depends on the ratio of the densities of the two liquids and on the elevations h_{A2} and h_T of the two overflow lines. Usually, the height h_{A2} is movable so that the interface level can be adjusted.

2.3   GENERAL MOLECULAR TRANSPORT EQUATION FOR MOMENTUM, HEAT, AND MASS TRANSFER

2.3A   General Molecular Transport Equation and General Property Balance

1. Introduction to transport processes.   In molecular transport processes in general we are concerned with the transfer or movement of a given property or entity by molecular movement through a system or medium which can be a fluid (gas or liquid) or a solid. This property that is being transferred can be mass, thermal energy (heat), or momentum. Each molecule of a system has a given quantity of the property mass, thermal energy, or momentum associated with it. When a difference in concentration of the property exists for any of these properties from one region to an adjacent region, a net transport of this property occurs. In dilute fluids such as gases, where the molecules are relatively far apart, the rate of transport of the property should be relatively fast, since few molecules are present to block the transport or interact. In dense fluids such as liquids, the molecules are close together, and transport or diffusion proceeds more slowly. The molecules in solids are even more closely packed than in liquids and molecular migration is even more restricted.

2. General molecular transport equation.   All three of the molecular transport processes of momentum, heat or thermal energy, and mass are characterized in the elementary sense by the same general type of transport equation. First we start by noting the following:

$$\text{rate of transfer process} = \frac{\text{driving force}}{\text{resistance}} \tag{2.3-1}$$

This states what is quite obvious—that we need a driving force to overcome a resistance in order to transport a property. This is similar to Ohm's law in electricity, where the rate of flow of electricity is proportional to the voltage drop (driving force) and inversely proportional to the resistance.

We can formalize Eq. (2.3-1) by writing an equation as follows for molecular transport or diffusion of a property:

$$\psi_z = -\delta \frac{d\Gamma}{dz} \tag{2.3-2}$$

where ψ_z is defined as the flux of the property as amount of property being transferred per unit time per unit cross-sectional area perpendicular to the z direction of flow in amount of property/s · m^2, δ is a proportionality constant called diffusivity in m^2/s, Γ is concentration of the property in amount of property/m^3, and z is the distance in the direction of flow in m.

If the process is at steady state, then the flux ψ_z is constant. Rearranging Eq. (2.3-2) and integrating,

$$\psi_z \int_{z_1}^{z_2} dz = -\delta \int_{\Gamma_1}^{\Gamma_2} d\Gamma \tag{2.3-3}$$

$$\psi_z = \frac{\delta(\Gamma_1 - \Gamma_2)}{z_2 - z_1} \tag{2.3-4}$$

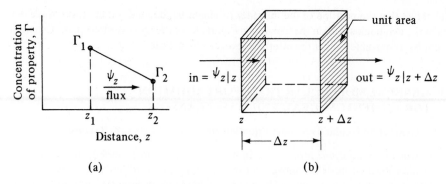

FIGURE 2.3-1. *Molecular transport of a property: (a) plot of concentration versus distance for steady state, (b) unsteady-state general property balance.*

A plot of the concentration Γ versus z is shown in Fig. 2.3-1a and is a straight line. Since the flux is in the direction 1 to 2 of decreasing concentration, the slope $d\Gamma/dz$ is negative, and the negative sign in Eq. (2.3-2) gives a positive flux in the direction 1 to 2. In Section 2.3B the specialized equations for momentum, heat, and mass transfer will be shown to be the same as Eq. (2.3-4) for the general property transfer.

EXAMPLE 2.3-1. Molecular Transport of a Property at Steady State

A property is being transported by diffusion through a fluid at steady state. At a given point 1 the concentration is 1.37×10^{-2} amount of property/m^3 and 0.72×10^{-2} at point 2 at a distance $z_2 = 0.40$ m. The diffusivity $\delta = 0.013$ m^2/s and the cross-sectional area is constant.

(a) Calculate the flux.
(b) Derive the equation for Γ as a function of distance.
(c) Calculate Γ at the midpoint of the path.

Solution: For part (a), substituting into Eq. (2.3-4),

$$\psi_z = \frac{\delta(\Gamma_1 - \Gamma_2)}{z_2 - z_1} = \frac{(0.013)(1.37 \times 10^{-2} - 0.72 \times 10^{-2})}{0.40 - 0}$$

$$= 2.113 \times 10^{-4} \text{ amount of property/s} \cdot m^2$$

For part (b), integrating Eq. (2.3-2) between Γ_1 and Γ and z_1 and z and rearranging,

$$\psi_z \int_{z_1}^{z} dz = -\delta \int_{\Gamma_1}^{\Gamma} d\Gamma \qquad \textbf{(2.3-5)}$$

$$\Gamma = \Gamma_1 + \frac{\psi_z}{\delta}(z_1 - z) \qquad \textbf{(2.3-6)}$$

For part (c), using the midpoint $z = 0.20$ m and substituting into Eq. (2.3-6),

$$\Gamma = 1.37 \times 10^{-2} + \frac{2.113 \times 10^{-4}}{0.013}(0 - 0.2)$$

$$= 1.045 \times 10^{-2} \text{ amount of property/}m^3$$

3. *General property balance for unsteady state.* In calculating the rates of transport in a system using the molecular transport equation (2.3-2), it is necessary to account for the amount of this property being transported in the entire system. This is done by writing a general property balance or conservation equation for the property (momentum, thermal energy, or mass) at unsteady state. We start by writing an equation for the z direction only, which accounts for all the property entering by molecular transport, leaving, being generated, and accumulating in a system shown in Fig. 2.3-1b, which is an element of volume $\Delta z(1)$ m^3 fixed in space.

$$\begin{pmatrix} \text{rate of} \\ \text{property in} \end{pmatrix} + \begin{pmatrix} \text{rate of generation} \\ \text{of property} \end{pmatrix}$$

$$= \begin{pmatrix} \text{rate of} \\ \text{property out} \end{pmatrix} + \begin{pmatrix} \text{rate of accumulation} \\ \text{of property} \end{pmatrix} \quad \textbf{(2.3-7)}$$

The rate of input is $(\psi_{z|z}) \cdot 1$ amount of property/s and the rate of output is $(\psi_{z|z+\Delta z}) \cdot 1$, where the cross-sectional area is 1.0 m^2. The rate of generation of the property is $R(\Delta z \cdot 1)$, where R is rate of generation of property/s$\cdot$m^3. The accumulation term is

$$\text{rate of accumulation of property} = \frac{\partial \Gamma}{\partial t}(\Delta z \cdot 1) \quad \textbf{(2.3-8)}$$

Substituting the various terms into Eq. (2.3-7),

$$(\psi_{z|z}) \cdot 1 + R(\Delta z \cdot 1) = (\psi_{z|z+\Delta z}) \cdot 1 + \frac{\partial \Gamma}{\partial t}(\Delta z \cdot 1) \quad \textbf{(2.3-9)}$$

Dividing by Δz and letting Δz go to zero,

$$\frac{\partial \Gamma}{\partial t} + \frac{\partial \psi_z}{\partial z} = R \quad \textbf{(2.3-10)}$$

Substituting Eq. (2.3-2) for ψ_z into (2.3-10) and assuming that δ is constant,

$$\frac{\partial \Gamma}{\partial t} - \delta \frac{\partial^2 \Gamma}{\partial z^2} = R \quad \textbf{(2.3-11)}$$

For the case where no generation is present,

$$\frac{\partial \Gamma}{\partial t} = \delta \frac{\partial^2 \Gamma}{\partial z^2} \quad \textbf{(2.3-12)}$$

This final equation relates the concentration of the property Γ to position z and time t.

Equations (2.3-11) and (2.3-12) are general equations for the conservation of momentum, thermal energy, or mass and will be used in many sections of this text. The equations consider here only molecular transport occurring and not other transport mechanisms such as convection and so on, which will be considered when the specific conservation equations are derived in later sections of this text for momentum, energy, and mass.

2.3B   Introduction to Molecular Transport

The kinetic theory of gases gives us a good physical interpretation of the motion of individual molecules in fluids. Because of their kinetic energy the molecules are in rapid random movement, often colliding with each other. Molecular transport or molecular diffusion of a property such as momentum, heat, or mass occurs in a fluid because of these random movements of individual molecules. Each individual molecule containing the property being transferred moves randomly in all directions, and there are fluxes in all directions. Hence, if there is a concentration gradient of the property, there will be a net flux of the property from high to low concentration. This occurs because equal numbers of molecules diffuse in each direction between the high-concentration and low-concentration regions.

1. Momentum transport and Newton's law. When a fluid is flowing in the x direction parallel to a solid surface, a velocity gradient exists where the velocity v_x in the x direction decreases as we approach the surface in the z direction. The fluid has x-directed momentum and its concentration is $v_x \rho$ momentum/m^3, where the momentum has units of kg $\cdot$ m/s. Hence, the units of $v_x \rho$ are (kg $\cdot$ m/s)/m^3. By random diffusion of molecules there is an exchange of molecules in the z direction, an equal number moving in each direction ($+z$ and $-z$ directions) between the faster-moving layer of molecules and the slower adjacent layer. Hence, the x-directed momentum has been transferred in the z direction from the faster- to the slower-moving layer. The equation for this transport of momentum is similar to Eq. (2.3-2) and is Newton's law of viscosity written as follows for constant density ρ:

$$\tau_{zx} = -\nu \frac{d(v_x \rho)}{dz} \tag{2.3-13}$$

where τ_{zx} is flux of x-directed momentum in the z direction, (kg $\cdot$ m/s)/s $\cdot$ m^2; ν is μ/ρ, the momentum diffusivity in m^2/s; z is the distance of transport or diffusion in m; ρ is the density in kg/m^3; and μ is the viscosity in kg/m $\cdot$ s.

2. Heat transport and Fourier's law. Fourier's law for molecular transport of heat or heat conduction in a fluid or solid can be written as follows for constant density ρ and heat capacity c_p:

$$\frac{q_z}{A} = -\alpha \frac{d(\rho c_p T)}{dz} \tag{2.3-14}$$

where q_z/A is the heat flux in J/s $\cdot$ m^2, α is the thermal diffusivity in m^2/s, and $\rho c_p T$ is the concentration of heat or thermal energy in J/m^3. When there is a temperature gradient in a fluid, equal numbers of molecules diffuse in each direction between the hot and the colder region. In this way energy is transferred in the z direction.

3. Mass transport and Fick's law. Fick's law for molecular transport of mass in a fluid or solid for constant total concentration in the fluid is

$$J_{Az}^* = -D_{AB} \frac{dc_A}{dz} \tag{2.3-15}$$

where J_{AZ}^* is the flux of A in kg mol A/s $\cdot$ m^2, D_{AB} is the molecular diffusivity of the molecule A in B in m^2/s, and c_A is the concentration of A in kg mol A/m^3. In a manner similar to momentum and heat transport, when there is a concentration gradient in a fluid, equal numbers of molecules diffuse in each direction between the high- and low-concentration regions and a net flux of mass occurs.

Hence, Eqs. (2.3-13), (2.3-14), and (2.3-15) for momentum, heat, and mass transfer are all similar to each other and to the general molecular transport equation (2.3-2). All equations have a flux on the left-hand side of each equation, a diffusivity in m²/s, and the derivative of the concentration with respect to distance. All three of the molecular transport equations are mathematically identical. Thus, we state that we have an analogy or similarity among them. It should be emphasized, however, that even though there is a mathematical analogy, the actual physical mechanisms occurring may be totally different. For example, in mass transfer two components are often being transported by relative motion through one another. In heat transport in a solid, the molecules are relatively stationary and the transport is done mainly by the electrons. Transport of momentum can occur by several types of mechanisms. More-detailed considerations of each of the transport processes of momentum, energy, and mass are presented in this and succeeding chapters.

2.4 VISCOSITY OF FLUIDS

2.4A Newton's Law and Viscosity

When a fluid is flowing through a closed channel such as a pipe or between two flat plates, either of two types of flow may occur, depending on the velocity of this fluid. At low velocities the fluid tends to flow without lateral mixing, and adjacent layers slide past one another like playing cards. There are no cross currents perpendicular to the direction of flow, nor eddies or swirls of fluid. This regime or type of flow is called *laminar flow*. At higher velocities eddies form, which leads to lateral mixing. This is called *turbulent flow*. The discussion in this section is limited to laminar flow.

A fluid can be distinguished from a solid in this discussion of viscosity by its behavior when subjected to a stress (force per unit area) or applied force. An elastic solid deforms by an amount proportional to the applied stress. However, a fluid, when subjected to a similar applied stress, will continue to deform, that is, to flow at a velocity that increases with increasing stress. A fluid exhibits resistance to this stress. Viscosity is that property of a fluid which gives rise to forces that resist the relative movement of adjacent layers in the fluid. These *viscous forces* arise from forces existing between the molecules in the fluid and are similar in character to the *shear forces* in solids.

The ideas above can be clarified by a more quantitative discussion of viscosity. In Fig. 2.4-1 a fluid is contained between two infinite (very long and very wide) parallel plates. Suppose that the bottom plate is moving parallel to the top plate and at a constant velocity Δv_z m/s faster relative to the top plate because of a steady force F newtons being applied. This force is called the *viscous drag,* and it arises from the viscous forces in the fluid. The plates are Δy m apart. Each layer of liquid moves in the z direction. The layer immediately adjacent to the bottom plate is carried along at the velocity of this plate. The layer just above is at a

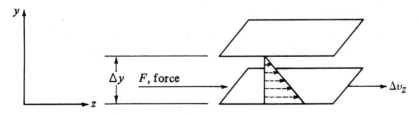

FIGURE 2.4-1.   *Fluid shear between two parallel plates.*

slightly slower velocity, each layer moving at a slower velocity as we go up in the y direction. This velocity profile is linear, with y direction as shown in Fig. 2.4-1. An analogy to a fluid is a deck of playing cards, where, if the bottom card is moved, all the other cards above will slide to some extent.

It has been found experimentally for many fluids that the force F in newtons is directly proportional to the velocity Δv_z in m/s and to the area A in m^2 of the plate used, and inversely proportional to the distance Δy in m. Or, as given by Newton's law of viscosity when the flow is laminar,

$$\frac{F}{A} = -\mu \frac{\Delta v_z}{\Delta y} \tag{2.4-1}$$

where μ is a proportionality constant called the *viscosity* of the fluid, in Pa $\cdot$ s or kg/m $\cdot$ s. If we let Δy approach zero, then, using the definition of the derivative,

$$\tau_{yz} = -\mu \frac{dv_z}{dy} \qquad \text{(SI units)} \tag{2.4-2}$$

where $\tau_{yz} = F/A$ and is the shear stress or force per unit area in newtons/m^2 (N/m^2). In the cgs system, F is in dynes, μ in g/cm $\cdot$ s, v_z in cm/s, and y in cm. We can also write Eq. (2.2-2) as

$$\tau_{yz} g_c = -\mu \frac{dv_z}{dy} \qquad \text{(English units)} \tag{2.4-3}$$

where τ_{yz} is in units of lb$_f$/ft^2.

The units of viscosity in the cgs system are g/cm $\cdot$ s, called *poise* or centipoise (cp). In the SI system, viscosity is given in Pa $\cdot$ s (N $\cdot$ s/m^2 or kg/m $\cdot$ s):

$$1 \text{ cp} = 1 \times 10^{-3} \text{ kg/m} \cdot \text{s} = 1 \times 10^{-3} \text{ Pa} \cdot \text{s} = 1 \times 10^{-3} \text{ N} \cdot \text{s/m}^2 \qquad \text{(SI)}$$

$$1 \text{ cp} = 0.01 \text{ poise} = 0.01 \text{ g/cm} \cdot \text{s}$$

$$1 \text{ cp} = 6.7197 \times 10^{-4} \text{ lb}_m/\text{ft} \cdot \text{s}$$

Other conversion factors for viscosity are given in Appendix A.1. Sometimes the viscosity is given as μ/ρ, kinematic viscosity, in m^2/s or cm^2/s, where ρ is the density of the fluid.

EXAMPLE 2.4-1. Calculation of Shear Stress in a Liquid
Referring to Fig. 2.4-1, the distance between plates is $\Delta y = 0.5$ cm, $\Delta v_z = 10$ cm/s, and the fluid is ethyl alcohol at 273 K having a viscosity of 1.77 cp (0.0177 g/cm $\cdot$ s).
  (a) Calculate the shear stress τ_{yz} and the velocity gradient or shear rate dv_z/dy using cgs units.
  (b) Repeat, using lb force, s, and ft units (English units).
  (c) Repeat, using SI units.

Solution: We can substitute directly into Eq. (2.4-1) or we can integrate Eq. (2.4-2). Using the latter method, rearranging Eq. (2.4-2), calling the bottom plate point 1, and integrating,

$$\tau_{yz} \int_{y_1=0}^{y_2=0.5} dy = -\mu \int_{v_1=10}^{v_2=0} dv_z \tag{2.4-4}$$

$$\tau_{yz} = \mu \frac{v_1 - v_2}{y_2 - y_1} \qquad \text{(2.4-5)}$$

Substituting the known values,

$$\tau_{yz} = \mu \frac{v_1 - v_2}{y_2 - y_1} = \left(0.0177 \frac{g}{cm \cdot s}\right) \frac{(10 - 0) \text{ cm/s}}{(0.5 - 0) \text{ cm}}$$

$$= 0.354 \frac{g \cdot cm/s^2}{cm^2} = 0.354 \frac{dyn}{cm^2} \qquad \text{(2.4-6)}$$

To calculate the shear rate dv_z/dy, since the velocity change is linear with y,

$$\text{shear rate} = \frac{dv_z}{dy} = \frac{\Delta v_z}{\Delta y} = \frac{(10 - 0) \text{ cm/s}}{(0.5 - 0) \text{ cm}} = 20.0 \text{ s}^{-1} \qquad \text{(2.4-7)}$$

For part (b), using lb force units and the viscosity conversion factor from Appendix A.1,

$$\mu = 1.77 \text{ cp}(6.7197 \times 10^{-4} \text{ lb}_m/\text{ft} \cdot \text{s})/\text{cp}$$

$$= 1.77(6.7197 \times 10^{-4}) \text{ lb}_m/\text{ft} \cdot \text{s}$$

Integrating Eq. (2.4-3),

$$\tau_{yz} = \frac{\mu \text{ lb}_m/\text{ft} \cdot \text{s}}{g_c \dfrac{\text{lb}_m \cdot \text{ft}}{\text{lb}_f \cdot \text{s}^2}} \frac{(v_1 - v_2)\text{ft/s}}{(y_2 - y_1) \text{ ft}} \qquad \text{(2.4-8)}$$

Substituting known values into Eq. (2.4-8) and converting Δv_z to ft/s and Δy to ft, $\tau_{yz} = 7.39 \times 10^{-4} \text{ lb}_f/\text{ft}^2$. Also, $dv_z/dy = 20 \text{ s}^{-1}$.

For part (c), $\Delta y = 0.5/100 = 0.005$ m, $\Delta v_z = 10/100 = 0.1$ m/s, and $\mu = 1.77 \times 10^{-3} \text{ kg/m} \cdot \text{s} = 1.77 \times 10^{-3} \text{ Pa} \cdot \text{s}$. Substituting into Eq. (2.4-5),

$$\tau_{yz} = (1.77 \times 10^{-3})(0.10)/0.005 = 0.0354 \text{ N/m}^2$$

The shear rate will be the same at 20.0 s^{-1}.

2.4B Momentum Transfer in a Fluid

The shear stress τ_{yz} in Eqs. (2.4-1)–(2.4-3) can also be interpreted as a *flux of z-directed momentum in the y direction,* which is the rate of flow of momentum per unit area. The units of momentum are mass times velocity in kg · m/s. The shear stress can be written

$$\tau_{yz} = \frac{\text{kg} \cdot \text{m/s}}{\text{m}^2 \cdot \text{s}} = \frac{\text{momentum}}{\text{m}^2 \cdot \text{s}} \qquad \text{(2.4-9)}$$

This gives an amount of momentum transferred per second per unit area.

This can be shown by considering the interaction between two adjacent layers of a fluid in Fig. 2.4-1 which have different velocities, and hence different momentum, in the z direction. The random motions of the molecules in the faster-moving layer send some of the molecules into the slower-moving layer, where they collide with the slower-moving molecules and tend to speed them up or increase their momentum in the z direction. Also, in the same fashion, molecules in the slower layer tend to retard those in the faster layer. This exchange of molecules between layers produces a transfer or flux of z-directed momentum from high-velocity to low-velocity layers. The negative sign in Eq. (2.4-2) indicates that momentum is transferred down the gradient from high- to low-velocity regions. This is similar to the transfer of heat from high- to low-temperature regions.

2.4C Viscosities of Newtonian Fluids

Fluids that follow Newton's law of viscosity, Eqs. (2.4-1)–(2.4-3), are called *Newtonian fluids*. For a Newtonian fluid, there is a linear relation between the shear stress τ_{yz} and the velocity gradient dv_z/dy (rate of shear). This means that the viscosity μ is a constant and is independent of the rate of shear. For non-Newtonian fluids, the relation between τ_{yz} and dv_z/dy is not linear; that is, the viscosity μ does not remain constant but is a function of shear rate. Certain liquids do not obey this simple law of Newton's. These are primarily pastes, slurries, high polymers, and emulsions. The science of the flow and deformation of fluids is often called *rheology*. A discussion of non-Newtonian fluids will not be given here but will be included in Section 3.5.

The viscosity of gases, which are Newtonian fluids, increases with temperature and is approximately independent of pressure up to a pressure of about 1000 kPa. At higher pressures, the viscosity of gases increases with increase in pressure. For example, the viscosity of N_2 gas at 298 K approximately doubles in going from 100 kPa to about 5×10^4 kPa (R1). In liquids, the viscosity decreases with increasing temperature. Since liquids are essentially incompressible, the viscosity is not affected by pressure.

In Table 2.4-1 some experimental viscosity data are given for some typical pure fluids at 101.32 kPa. The viscosities for gases are the lowest and do not differ markedly from gas to gas, being about 5×10^{-6} to 3×10^{-5} Pa·s. The viscosities for liquids are much greater. The value for water at 293 K is about 1×10^{-3} and for glycerol 1.069 Pa·s. Hence, there are great differences between viscosities of liquids. More complete tables of viscosities are given for water in Appendix A.2, for inorganic and organic liquids and gases in Appendix A.3, and for biological and food liquids in Appendix A.4. Extensive data are available in other references (P1, R1, W1, L1). Methods of estimating viscosities of gases and liquids when experimental

TABLE 2.4-1. *Viscosities of Some Gases and Liquids at 101.32 kPa Pressure*

	Gases				Liquids		
Substance	Temp., K	Viscosity (Pa·s) 10^3 or (kg/m·s) 10^3	Ref.	Substance	Temp., K	Viscosity (Pa·s) 10^3 or (kg/m·s) 10^3	Ref.
Air	293	0.01813	N1	Water	293	1.0019	S1
CO_2	273	0.01370	R1		373	0.2821	S1
	373	0.01828	R1	Benzene	278	0.826	R1
CH_4	293	0.01089	R1				
				Glycerol	293	1069	L1
SO_2	373	0.01630	R1	Hg	293	1.55	R2
				Olive oil	303	84	E1

data are not available are summarized elsewhere (R1). These estimation methods for gases at pressures below 100 kPa are reasonably accurate, with an error within about ±5%, but the methods for liquids are often quite inaccurate.

2.5 TYPES OF FLUID FLOW AND REYNOLDS NUMBER

2.5A Introduction and Types of Fluid Flow

The principles of the statics of fluids, treated in Section 2.2, are almost an exact science. On the other hand, the principles of the motions of fluids are quite complex. The basic relations describing the motions of a fluid are the equations for the overall balances of mass, energy, and momentum, which will be covered in the following sections.

These overall or macroscopic balances will be applied to a finite enclosure or control volume fixed in space. We use the term "overall" because we wish to describe these balances from outside the enclosure. The changes inside the enclosure are determined in terms of the properties of the streams entering and leaving and the exchanges of energy between the enclosure and its surroundings.

When making overall balances on mass, energy, and momentum we are not interested in the details of what occurs inside the enclosure. For example, in an overall balance, average inlet and outlet velocities are considered. However, in a differential balance, the velocity distribution inside an enclosure can be obtained by the use of Newton's law of viscosity.

In this section we first discuss the two types of fluid flow that can occur: laminar flow and turbulent flow. Also, the Reynolds number used to characterize the regimes of flow is considered. Then in Sections 2.6, 2.7, and 2.8, the overall mass balance, energy balance, and momentum balance are covered together with a number of applications. Finally, a discussion is given in Section 2.9 on the methods of making a shell balance on an element to obtain the velocity distribution in the element and the pressure drop.

2.5B Laminar and Turbulent Flow

The type of flow occurring in a fluid in a channel is important in fluid dynamics problems. When fluids move through a closed channel of any cross section, either of two distinct types of flow can be observed, according to the conditions present. These two types of flow can commonly be seen in a flowing open stream or river. When the velocity of flow is slow, the flow patterns are smooth. However, when the velocity is quite high, an unstable pattern is observed, in which eddies or small packets of fluid particles are present, moving in all directions and at all angles to the normal line of flow.

The first type of flow, at low velocities, where the layers of fluid seem to slide by one another without eddies or swirls being present, is called *laminar flow,* and Newton's law of viscosity holds, as discussed in Section 2.4A. The second type of flow, at higher velocities, where eddies are present giving the fluid a fluctuating nature, is called *turbulent flow.*

The existence of laminar and turbulent flow is most easily visualized by the experiments of Reynolds. His experiments are shown in Fig. 2.5-1. Water was allowed to flow at steady state through a transparent pipe with the flow rate controlled by a valve at the end of the pipe. A fine, steady stream of dyed water was introduced from a fine jet as shown and its flow pattern observed. At low rates of water flow, the dye pattern was regular and formed a single line or stream similar to a thread, as shown in Fig. 2.5-1a. There was no lateral mixing of the fluid, and it flowed in streamlines down the tube. By putting in additional jets at other points in the

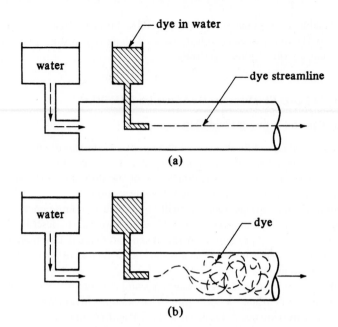

FIGURE 2.5-1. *Reynolds' experiment for different types of flow:*
(a) laminar flow; (b) turbulent flow.

pipe cross section, it was shown that there was no mixing in any parts of the tube and the fluid flowed in straight parallel lines. This type of flow is called laminar or *viscous flow*.

As the velocity was increased, it was found that at a definite velocity the thread of dye became dispersed and the pattern was very erratic, as shown in Fig. 2.5-1b. This type of flow is known as turbulent flow. The velocity at which the flow changes is known as the *critical velocity*.

2.5C   Reynolds Number

Studies have shown that the transition from laminar to turbulent flow in tubes is not only a function of velocity but also of density and viscosity of the fluid and the tube diameter. These variables are combined into the Reynolds number, which is dimensionless:

$$N_{Re} = \frac{Dv\rho}{\mu} \tag{2.5-1}$$

where N_{Re} is the Reynolds number, D the diameter in m, ρ the fluid density in kg/m^3, μ the fluid viscosity in Pa · s, and v the average velocity of the fluid in m/s (where average velocity is defined as the volumetric rate of flow divided by the cross-sectional area of the pipe). Units in the cgs system are D in cm, ρ in g/cm^3, μ in g/cm · s, and v in cm/s. In the English system D is in ft, ρ in lb$_m$/ft^3, μ in lb$_m$/ft · s, and v in ft/s.

The instability of the flow that leads to disturbed or turbulent flow is determined by the ratio of the kinetic or inertial forces to the viscous forces in the fluid stream. The inertial forces are proportional to ρv^2 and the viscous forces to $\mu v/D$, and the ratio $\rho v^2/(\mu v/D)$ is the Reynolds number $Dv\rho/\mu$. Further explanation and derivation of dimensionless numbers are given in Section 3.11.

For a straight circular pipe, when the value of the Reynolds number is less than 2100, the flow is always laminar. When the value is over 4000, the flow will be turbulent, except in very special cases. In between—called the *transition region*—the flow can be viscous or turbulent, depending upon the apparatus details, which cannot be predicted.

EXAMPLE 2.5-1   Reynolds Number in a Pipe

Water at 303 K is flowing at the rate of 10 gal/min in a pipe having an inside diameter (ID) of 2.067 in. Calculate the Reynolds number using both English units and SI units.

Solution: From Appendix A.1, 7.481 gal = 1 ft^3. The flow rate is calculated as

$$\text{flow rate} = \left(10.0\,\frac{\text{gal}}{\text{min}}\right)\left(\frac{1\,\text{ft}^3}{7.481\,\text{gal}}\right)\left(\frac{1\,\text{min}}{60\,\text{s}}\right)$$

$$= 0.0223\,\text{ft}^3/\text{s}$$

$$\text{pipe diameter, } D = \frac{2.067}{12} = 0.172\,\text{ft}$$

$$\text{cross-sectional area of pipe} = \frac{\pi D^2}{4} = \frac{\pi(0.172)^2}{4} = 0.0233\,\text{ft}^2$$

$$\text{velocity in pipe, } v = \left(0.0223\,\frac{\text{ft}^3}{\text{s}}\right)\left(\frac{1}{0.0233\,\text{ft}^2}\right) = 0.957\,\text{ft/s}$$

From Appendix A.2, for water at 303 K (30°C),

$$\text{density, } \rho = 0.996(62.43)\,\text{lb}_\text{m}/\text{ft}^3$$

$$\text{viscosity, } \mu = (0.8007\,\text{cp})\left(6.7197 \times 10^{-4}\frac{\text{lb}_\text{m}/\text{ft} \cdot \text{s}}{\text{cp}}\right)$$

$$= 5.38 \times 10^{-4}\,\text{lb}_\text{m}/\text{ft} \cdot \text{s}$$

Substituting into Eq. (2.5-1).

$$N_\text{Re} = \frac{Dv\rho}{\mu} = \frac{(0.172\,\text{ft})(0.957\,\text{ft/s})(0.996 \times 62.43\,\text{lb}_\text{m}/\text{ft}^3)}{5.38 \times 10^{-4}\,\text{lb}_\text{m}/\text{ft} \cdot \text{s}}$$

$$= 1.905 \times 10^4$$

Hence, the flow is turbulent. Using SI units,

$$\rho = (0.996)(1000\,\text{kg/m}^3) = 996\,\text{kg/m}^3$$

$$D = (2.067\,\text{in.})(1\,\text{ft/12 in.})(1\,\text{m/3.2808 ft}) = 0.0525\,\text{m}$$

$$v = \left(0.957\,\frac{\text{ft}}{\text{s}}\right)(1\,\text{m/3.2808 ft}) = 0.2917\,\text{m/s}$$

$$\mu = (0.8007\,\text{cp})\left(1 \times 10^{-3}\frac{\text{kg/m} \cdot \text{s}}{\text{cp}}\right) = 8.007 \times 10^{-4}\frac{\text{kg}}{\text{m} \cdot \text{s}}$$

$$= 8.007 \times 10^{-4}\,\text{Pa} \cdot \text{s}$$

$$N_\text{Re} = \frac{Dv\rho}{\mu} = \frac{(0.0525\,\text{m})(0.2917\,\text{m/s})(996\,\text{kg/m}^3)}{8.007 \times 10^{-4}\,\text{kg/m} \cdot \text{s}} = 1.905 \times 10^4$$

2.6   OVERALL MASS BALANCE AND CONTINUITY EQUATION

2.6A   Introduction and Simple Mass Balances

In fluid dynamics fluids are in motion. Generally, they are moved from place to place by means of mechanical devices such as pumps or blowers, by gravity head, or by pressure, and flow through systems of piping and/or process equipment. The first step in the solution of flow problems is generally to apply the principles of the conservation of mass to the whole system or to any part of the system. First, we will consider an elementary balance on a simple geometry, and later we shall derive the general mass-balance equation.

Simple mass or material balances were introduced in Section 1.5, where

$$\text{input} = \text{output} + \text{accumulation} \qquad \textbf{(1.5-1)}$$

Since, in fluid flow, we are usually working with rates of flow and usually at steady state, the rate of accumulation is zero and we obtain

$$\text{rate of input} = \text{rate of output (steady state)} \qquad \textbf{(2.6-1)}$$

In Fig. 2.6-1 a simple flow system is shown, where fluid enters section 1 with an average velocity v_1 m/s and density ρ_1 kg/m^3. The cross-sectional area is A_1 m^2. The fluid leaves section 2 with average velocity v_2. The mass balance, Eq. (2.6-1), becomes

$$m = \rho_1 A_1 v_1 = \rho_2 A_2 v_2 \qquad \textbf{(2.6-2)}$$

where m = kg/s. Often, $v\rho$ is expressed as $G = v\rho$, where G is mass velocity or mass flux in kg/s $\cdot$ m^2. In English units, v is in ft/s, ρ in lb$_m$/ft^3, A in ft^2, m in lb$_m$/s, and G in lb$_m$/s $\cdot$ ft^2.

EXAMPLE 2.6-1.   Flow of Crude Oil and Mass Balance
A petroleum crude oil having a density of 892 kg/m^3 is flowing through the piping arrangement shown in Fig. 2.6-2 at a total rate of 1.388×10^{-3} m^3/s entering pipe 1.

The flow divides equally in each of pipes 3. The steel pipes are schedule 40 pipe (see Appendix A.5 for actual dimensions). Calculate the following, using SI units:

    (a) The total mass flow rate m in pipe 1 and pipes 3.
    (b) The average velocity v in 1 and 3.
    (c) The mass velocity G in 1.

Solution: From Appendix A.5, the dimensions of the pipes are as follows: 2-in. pipe: D_1 (ID) = 2.067 in.; cross-sectional area

$$A_1 = 0.02330 \text{ ft}^2 = 0.02330(0.0929) = 2.165 \times 10^{-3} \text{ m}^2$$

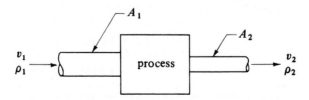

FIGURE 2.6-1.   *Mass balance on flow system.*

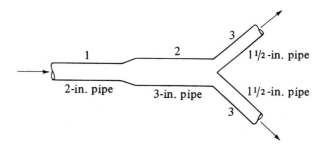

FIGURE 2.6-2. *Piping arrangement for Example 2.6-1.*

$1\frac{1}{2}$-in. pipe: D_3 (ID) = 1.610 in.; cross-sectional area

$$A_3 = 0.01414 \text{ ft}^2 = 0.01414(0.0929) = 1.313 \times 10^{-3} \text{ m}^2$$

The total mass flow rate is the same through pipes 1 and 2 and is

$$m_1 = (1.388 \times 10^{-3} \text{ m}^3/\text{s})(892 \text{ kg/m}^3) = 1.238 \text{ kg/s}$$

Since the flow divides equally in each of pipes 3,

$$m_3 = \frac{m_1}{2} = \frac{1.238}{2} = 0.619 \text{ kg/s}$$

For part (b), using Eq. (2.6-2) and solving for v,

$$v_1 = \frac{m_1}{\rho_1 A_1} = \frac{1.238 \text{ kg/s}}{(892 \text{ kg/m}^3)(2.165 \times 10^{-3} \text{ m}^2)} = 0.641 \text{ m/s}$$

$$v_3 = \frac{m_3}{\rho_3 A_3} = \frac{0.619}{(892)(1.313 \times 10^{-3})} = 0.528 \text{ m/s}$$

For part (c),

$$G_1 = v_1 \rho_1 = \frac{m_1}{A_1} = \frac{1.238}{2.165 \times 10^{-3}} = 572 \frac{\text{kg}}{\text{s} \cdot \text{m}^2}$$

2.6B Control Volume for Balances

The laws for the conservation of mass, energy, and momentum are all stated in terms of a system; these laws give the interaction of a system with its surroundings. A *system* is defined as a collection of fluid of fixed identity. However, in flow of fluids, individual particles are not easily identifiable. As a result, attention is focused on a given space through which the fluid flows rather than on a given mass of fluid. The method used, which is more convenient, is to select a control volume, which is a region fixed in space through which the fluid flows.

In Fig. 2.6-3 the case of a fluid flowing through a conduit is shown. The control surface shown as a dashed line is the surface surrounding the control volume. In most problems part of the control surface will coincide with some boundary, such as the wall of the conduit. The remaining part of the control surface is a hypothetical surface through which the fluid can

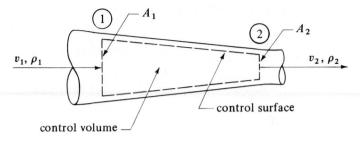

FIGURE 2.6-3. *Control volume for flow through a conduit.*

flow, shown as point 1 and point 2 in Fig. 2.6-3. The control-volume representation is analogous to the open system of thermodynamics.

2.6C Overall Mass-Balance Equation

In deriving the general equation for the overall balance of the property mass, the law of conservation of mass may be stated as follows for a control volume where no mass is being generated.

$$\left(\begin{array}{l}\text{rate of mass output}\\\text{from control volume}\end{array}\right) - \left(\begin{array}{l}\text{rate of mass input}\\\text{from control volume}\end{array}\right)$$

$$+ \left(\begin{array}{l}\text{rate of mass accumulation}\\\text{in control volume}\end{array}\right) = 0 \quad (\text{rate of mass generation}) \quad \textbf{(2.6-3)}$$

We now consider the general control volume fixed in space and located in a fluid flow field, as shown in Fig. 2.6-4. For a small element of area dA m^2 on the control surface, the rate of mass efflux from this element $= (\rho v)(dA \cos \alpha)$, where $(dA \cos \alpha)$ is the area dA projected in a direction normal to the velocity vector v, α is the angle between the velocity vector v and the outward-directed unit normal vector n to dA, and ρ is the density in kg/m^3. The quantity ρv has units of kg/s $\cdot$ m^2 and is called *a flux* or *mass velocity G.*

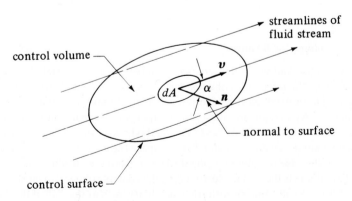

FIGURE 2.6-4. *Flow through a differential area dA on a control surface.*

From vector algebra we recognize that $(\rho v)(dA \cos \alpha)$ is the scalar or dot product $\rho(\boldsymbol{v} \cdot \boldsymbol{n}) \, dA$. If we now integrate this quantity over the entire control surface A, we have the net outflow of mass across the control surface, or the net mass efflux in kg/s from the entire control volume V:

$$\begin{pmatrix} \text{net mass efflux} \\ \text{from control volume} \end{pmatrix} = \iint_A v\rho \cos \alpha \, dA = \iint_A \rho(\boldsymbol{v} \cdot \boldsymbol{n}) \, dA \qquad \textbf{(2.6-4)}$$

We should note that if mass is entering the control volume, that is, flowing inward across the control surface, the next efflux of mass in Eq. (2.6-4) is negative, since $\alpha > 90°$ and $\cos \alpha$ is negative. Hence, there is a net influx of mass. If $\alpha < 90°$, there is a net efflux of mass.

The rate of accumulation of mass within the control volume V can be expressed as follows:

$$\begin{pmatrix} \text{rate of mass accumulation} \\ \text{in control volume} \end{pmatrix} = \frac{\partial}{\partial t} \iiint_V \rho \, dV = \frac{dM}{dt} \qquad \textbf{(2.6-5)}$$

where M is the mass of fluid in the volume in kg. Substituting Eqs. (2.6-4) and (2.6-5) into (2.6-3), we obtain the general form of the overall mass balance:

$$\iint_A \rho(\boldsymbol{v} \cdot \boldsymbol{n}) \, dA + \frac{\partial}{\partial t} \iiint_V \rho \, dV = 0 \qquad \textbf{(2.6-6)}$$

The use of Eq. (2.6-6) can be shown for a common situation of steady-state one-dimensional flow, where all the flow inward is normal to A_1 and outward normal to A_2, as shown in Fig. 2.6-3. When the velocity v_2 leaving (Fig. 2.6-3) is normal to A_2, the angle α_2 between the normal to the control surface and the direction of the velocity is $0°$, and $\cos \alpha_2 = 1.0$. Where v_1 is directed inward, $\alpha_1 > \pi/2$, and for the case in Fig. 2.6-3, α_1 is $180°$ ($\cos \alpha_1 = -1.0$). Since α_2 is $0°$ and α_1 is $180°$, using Eq. (2.6-4),

$$\iint_A v\rho \cos \alpha \, dA = \iint_{A_2} v\rho \cos \alpha_2 \, dA + \iint_{A_1} v\rho \cos \alpha_1 \, dA \qquad \textbf{(2.6-7)}$$

$$= v_2 \rho_2 A_2 - v_1 \rho_1 A_1$$

For steady state, $dM/dt = 0$ in Eq. (2.6-5), and Eq. (2.6-6) becomes

$$m = \rho_1 v_1 A_1 = \rho_2 v_2 A_2 \qquad \textbf{(2.6-2)}$$

which is Eq. (2.6-2), derived earlier.

In Fig. 2.6-3 and Eqs. (2.6-3)–(2.6-7) we were not concerned with the composition of any of the streams. These equations can easily be extended to represent an overall mass balance for component i in a multicomponent system. For the case shown in Fig. 2.6-3, we combine Eqs. (2.6-5), (2.6-6), and (2.6-7), add a generation term, and obtain

$$m_{i2} - m_{i1} + \frac{dM_i}{dt} = R_i \qquad \textbf{(2.6-8)}$$

where m_{i2} is the mass flow rate of component i leaving the control volume and R_i is the rate of generation of component i in the control volume in kg per unit time. (Diffusion fluxes are neglected here or are assumed negligible.) In some cases, of course, $R_i = 0$ for no generation. Often it is more convenient to use Eq. (2.6-8) written in molar units.

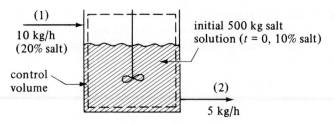

FIGURE 2.6-5. *Control volume for flow in a stirred tank for Example 2.6-2.*

EXAMPLE 2.6-2. *Overall Mass Balance in Stirred Tank*

Initially, a tank holds 500 kg of salt solution containing 10% salt. At point (1) in the control volume in Fig. 2.6-5, a stream enters at a constant flow rate of 10 kg/h containing 20% salt. A stream leaves at point (2) at a constant rate of 5 kg/h. The tank is well stirred. Derive an equation relating the weight fraction w_A of the salt in the tank at any time t in hours.

Solution: First we make a total mass balance using Eq. (2.6-7) for the net total mass efflux from the control volume:

$$\iint_A v\rho \cos \alpha \, dA = m_2 - m_1 = 5 - 10 = -5 \text{ kg solution/h} \qquad \textbf{(2.6-9)}$$

From Eq. (2.6-5), where M is total kg of solution in control volume at time t,

$$\frac{\partial}{\partial t} \iiint_V \rho \, dV = \frac{dM}{dt} \qquad \textbf{(2.6-5)}$$

Substituting Eqs. (2.6-5) and (2.6-9) into (2.6-6), and then integrating,

$$-5 + \frac{dM}{dt} = 0$$

$$\int_{M=500}^{M} dM = 5 \int_{t=0}^{t} dt \qquad \textbf{(2.6-10)}$$

$$M = 5t + 500 \qquad \textbf{(2.6-11)}$$

Equation (2.6-11) relates the total mass M in the tank at any time to t.

Next, making a component A salt balance, let w_A = weight fraction of salt in tank at time t and also the concentration in the stream m_2 leaving at time t. Again using Eq. (2.6-7) but for a salt balance,

$$\iint_A v\rho \cos \alpha \, dA = (5)w_A - 10(0.20) = 5w_A - 2 \text{ kg salt/h} \qquad \textbf{(2.6-12)}$$

Using Eq. (2.6-5) for a salt balance,

$$\frac{\partial}{\partial t} \iiint_V \rho \, dV = \frac{d}{dt}(Mw_A) = \frac{M \, dw_A}{dt} + w_A \frac{dM}{dt} \text{ kg salt/h} \qquad \textbf{(2.6-13)}$$

Substituting Eqs. (2.6-12) and (2.6-13) into (2.6-6),

$$5w_A - 2 + M\frac{dw_A}{dt} + w_A\frac{dM}{dt} = 0 \qquad \textbf{(2.6-14)}$$

Substituting the value for M from Eq. (2.6-11) into (2.6-14), separating variables, integrating, and solving for w_A,

$$5w_A - 2 + (500 + 5t)\frac{dw_A}{dt} + w_A\frac{d(500 + 5t)}{dt} = 0$$

$$5w_A - 2 + (500 + 5t)\frac{dw_A}{dt} + 5w_A = 0$$

$$\int_{w_A=0.10}^{w_A}\frac{dw_A}{2 - 10w_A} = \int_{t=0}^{t}\frac{dt}{500 + 5t}$$

$$-\frac{1}{10}\ln\left(\frac{2 - 10w_A}{1}\right) = \frac{1}{5}\ln\left(\frac{500 + 5t}{500}\right) \qquad \textbf{(2.6-15)}$$

$$w_A = -0.1\left(\frac{100}{100 + t}\right)^2 + 0.20 \qquad \textbf{(2.6-16)}$$

Note that Eq. (2.6-8) for component i could have been used for the salt balance with $R_i = 0$ (no generation).

2.6D   Average Velocity to Use in Overall Mass Balance

In solving the case in Eq. (2.6-7), we assumed a constant velocity v_1 at section 1 and constant v_2 at section 2. If the velocity is not constant but varies across the surface area, an average or bulk velocity is defined by

$$v_{av} = \frac{1}{A}\iint_A v\,dA \qquad \textbf{(2.6-17)}$$

for a surface over which v is normal to A and the density ρ is assumed constant.

EXAMPLE 2.6-3.   Variation of Velocity Across Control Surface and Average Velocity

For the case of incompressible flow (ρ is constant) through a circular pipe of radius R, the velocity profile is parabolic for laminar flow as follows:

$$v = v_{max}\left[1 - \left(\frac{r}{R}\right)^2\right] \qquad \textbf{(2.6-18)}$$

where v_{max} is the maximum velocity at the center where $r = 0$ and v is the velocity at a radial distance r from the center. Derive an expression for the average or bulk velocity v_{av} to use in the overall mass-balance equation.

Solution: The average velocity is represented by Eq. (2.6-17). In Cartesian coordinates dA is $dx\,dy$. However, using polar coordinates, which are more appropriate for a pipe, $dA = r\,dr\,d\theta$, where θ is the angle in polar coordinates. Substituting Eq. (2.6-18), $dA = r\,dr\,d\theta$, and $A = \pi R^2$ into Eq. (2.6-17) and integrating,

$$
v_{av} = \frac{1}{\pi R^2} \int_0^{2\pi} \int_0^R v_{max}\left[1 - \left(\frac{r}{R}\right)^2\right] r\,dr\,d\theta
$$

$$
= \frac{v_{max}}{\pi R^4} \int_0^{2\pi} \int_0^R (R^2 - r^2)r\,dr\,d\theta
$$

$$
= \frac{v_{max}}{\pi R^4}(2\pi - 0)\left(\frac{R^4}{2} - \frac{R^4}{4}\right) \tag{2.6-19}
$$

$$
v_{av} = \frac{v_{max}}{2} \tag{2.6-20}
$$

In this discussion overall or macroscopic mass balances were made because we wish to describe these balances from outside the enclosure. In this section on overall mass balances, some of the equations presented may have seemed quite obvious. However, the purpose was to develop the methods which should be helpful in the next sections. Overall balances will also be made on energy and momentum in the next sections. These overall balances do not tell us the details of what happens inside. However, in Section 2.9 a shell momentum balance will be made in order to obtain these details, which will give us the velocity distribution and pressure drop. To further study these details of the processes occurring inside the enclosure, differential balances rather than shell balances can be written; these are discussed later in Sections 3.6–3.9 on differential equations of continuity and momentum transfer, Sections 5.6 and 5.7 on differential equations of energy change and boundary-layer flow, and Section 7.5B on differential equations of continuity for a binary mixture.

2.7   OVERALL ENERGY BALANCE

2.7A   Introduction

The second property to be considered in the overall balances on a control volume is energy. We shall apply the principle of the conservation of energy to a control volume fixed in space in much the same manner as the principle of conservation of mass was used to obtain the overall mass balance. The energy-conservation equation will then be combined with the first law of thermodynamics to obtain the final overall energy-balance equation.

We can write the first law of thermodynamics as

$$
\Delta E = Q - W \tag{2.7-1}
$$

where E is the total energy per unit mass of fluid, Q is the heat *absorbed* per unit mass of fluid, and W is the work of all kinds done per unit mass of fluid *upon* the surroundings. In the calculations, each term in the equation must be expressed in the same type of units, such as J/kg (SI), btu/lb$_m$, or ft · lb$_f$/lb$_m$ (English).

Since mass carries with it associated energy due to its position, motion, or physical state,